Edexcel

Biology A2
Student Workbook

Model Answers: 2011

This model answer booklet is a companion publication to provide answers for the exercises in the Edexcel Biology A2 Student Workbook 2011 edition. These answers have been produced as a separate publication to keep the cost of the workbook itself to a minimum, as well as to prevent easy access to the answers by students. In most cases, simply the answer is given with no working or calculations described. A few, however, have been provided with greater detail because of their more difficult nature.

www.biozone.co.uk

ISBN 978-1-877462-57-3

PHOTOCOPYING PROHIBITED

including photocopying under a photocopy licence scheme such as CAL

BIOZONE

Additional copies of this Model Answers book may be purchased directly from the publisher

UK & EUROPE:
BIOZONE Learning Media (UK) Ltd.
Bretby Business Park, Ashby Road, Bretby,
Burton upon Trent, DE15 0YZ, **UK**
Telephone: 01283-553-257
Fax: 01283-553-258
Email: sales@biozone.co.uk

AUSTRALIA:
BIOZONE Learning Media Australia
P.O. Box 2841, Burleigh BC,
QLD 4220, **Australia**
Telephone: +61 7-5535-4896
Fax: +61 7-5508-2432
Email: sales@biozone.com.au

REST OF WORLD:
BIOZONE International Ltd.
P.O. Box 13-034,
Hamilton 3251, **New Zealand**
Telephone: +64 7-856-8104
Fax: +64 7-856-9243
Email: sales@biozone.co.nz

Contents

Contents

Contents

Drug Development

Hypotheses and Predictions (page 13)

1. Prediction: Woodlice are more likely to be found in moist habitats than in dry habitats.

2. (a) **Bacterial cultures**:
   Prediction: Bacterial strain A will grow more rapidly at 37°C than at room temperature (19°C).
   Outline of the investigation: Set up agar plates of bacterial strain A, using the streak plating method. Place 4 plates in a 37°C incubator and 4 on the lab bench. Leave all 8 plates for the same length of time (e.g. 24 hours), with all other conditions identical. Measure the coverage of the agar plates with bacteria (as a percentage).
   (b) **Plant cloning**:
   Prediction: A greater concentration of hormone A increases the rate of root growth in plant A.
   Outline of the investigation: Set up 6 agar plates infused with increasing concentrations of hormone A (e.g. 1 mgl^{-1}, 5 mgl^{-1}, 10 mgl^{-1}, 50 mgl^{-1}, 100 mgl^{-1}, 500 mgl^{-1}), and each plate with 12 clones of plant A. Measure root length each day for 20 days.

Choosing Your Topic (page 15)

There is no solution to this exercise. Students may use the space provided to develop any ideas they have for topics and identify the variables involved. Students should attempt to deal with quantitative data, at least for their response (dependent) variable, since these are more easily analysed meaningfully, using descriptive and inferential statistics.

Planning an Investigation (page 16)

1. Aim: To investigate the effect of temperature on the rate of catalase activity.

2. Hypothesis: The rate of catalase activity is dependent on temperature.

3. (a) Independent variable: Temperature.
   (b) Values: 10-60°C in uneven steps: 10°C, 20°C, 30°C, 60°C.
   (c) Unit: °C
   (d) Equipment: A means to maintain the test-tubes at the set temperatures, e.g. water baths. Equilibrate all reactants to the required temperatures in each case, before adding enzyme to the reaction tubes.

4. (a) Dependent variable: Height of oxygen bubbles.
   (b) Unit: mm
   (c) Equipment: Ruler; place vertically alongside the tube and read off the height (directly facing).

5. (a) Each temperature represents a treatment.
   (b) No. of tubes at each temperature = 2
   (c) Sample size: for each treatment = 2
   (d) Times the investigation repeated = 3

6. It would have been desirable to have had an extra tube with no enzyme to determine whether or not any oxygen was produced in the absence of enzyme.

7. Variables that might have been controlled (a-c):
   (a) Catalase from the same batch source and with the same storage history. Likewise for the H_2O_2. Storage and batch history can be determined.
   (b) Equipment of the same type and size (i.e. using test-tubes of the same dimensions, as well as volume). This could be checked before starting.
   (c) Same person doing the measurements of height each time. This should be decided beforehand.

   Note that some variables were controlled: The test-tube volume, and the volume of each reactant. Control of measurement error is probably the most important after these considerations.

8. Controlled variables should be monitored carefully to ensure that the only variable that changes between treatments (apart from the biological response) is the independent (manipulated) variable.

Asking Questions in Field Science (page 18)

1. Placing ponga logs into urban streams improves habitat and provides cover to promote the growth and reproduction of existing fish populations. Introducing farmed native fish would further enhance the biological diversity of the urban stream.

Designing Your Field Study (page 19)

1. (a) An appropriately sized sampling unit enables you to obtain sufficient data to test your predictions, but not so much that its analysis is too time consuming.
   (b) Recognising assumptions is critical to asking the appropriate questions in a study. It also allows you to recognise possible reasons for findings that do not support your predictions. **Note**: For even the simplest of studies, some background information is important. Making some intelligent assumptions based on present knowledge allows you to focus on the questions that you really want to answer.
   (c) Appropriate consideration of the environment helps to ensure that the environment is not significantly changed by your sampling activities.
   (d) Returning organisms to the same place after removal ensures that the usual distribution of organisms in the environment is preserved.
   (e) The sampling area must be large enough to incorporate sites representing the variations in habitat seen within the region for which predictions are being made (e.g. coniferous forest).

2. To test quadrat size it would be necessary to sample using a series of quadrats of increasing size. When only one species is involved, it is simplest to record the number of individuals for a range of different quadrat sizes. This gives some idea of variation in numbers and allows you to chose a quadrat size where the number recorded meets the needs of the data analysis (usually 10-100 individuals). **Note**: For community studies involving more than one species, the cumulative number of species recorded after each successive increase in quadrat size could be plotted (i.e. number of species vs quadrat size). Optimum quadrat size occurs when the number of species recorded stops increasing. A smaller size may be acceptable if you are prepared to record only dominant species. This always carries a risk that differences between areas may be missed.
   Checklist to be completed by the student.

Investigating Habitat Preference (page 21)

1. The major features of the habitat were given the numbers 1-5 so that a numerical scale was created

based on the degree of disturbance in the habitat.

2. Only cover types taking up more than 20% of the habitat were used as descriptors.

3. Fish were trapped over a long period of time in many areas that covered more than the known range of mudfish. In each area, traps were set at the same time of the day over a wide area, in a uniform way.

4. Mudfish weight and length were measured.

5. Mudfish prefer minimally disturbed environment.

6. Because few areas were found to have a disturbance rating of 1 or 2 the preferences there were too unreliable to distinguish between them.

Designing Your Experiment (page 22)

1. (a) Growing just one plant per pot removes the confounding effects of competition between plants (this would occur if plants were grown together).
   (b) The physical layout minimises any effects due to position of the pots on the bench. No one treatment is always near the edge or in the centre. Physical layout can affect the outcome of experimental treatments, especially those involving growth responses in plants. Physical conditions may vary considerably with different placements along a lab bench (near the window vs central), so treatments should be arranged so that all treatments, on average, are similarly affected.

2. The best way to account for natural variability between individuals is to increase the number of individuals you use in each treatment (sample size, n).

3. (a) Replication accounts for unforeseen effects in your set-up, i.e. allows you to evaluate the effects of **nuisance** variables over which you have no control.
   (b) Replication is limited by the amount of equipment, space, and/or time you have available.
   (c) You can compensate for a lack of replication by repeating the entire experiment several more times. This may be practicable only if the investigation does not involve lengthy growth responses.

   Checklist to be completed by the student.

Transforming Raw Data (page 24)

1. (a) Transforming data involves performing calculations using the raw data to determine such properties as rates, percentages, and totals.
   (b) The purpose of data transformation is to convert raw data into a more useful form.

2. (a) **Transformation**: Percentage (percentage cover)
   Reason: Abundance alone might not reflect the importance of a species in terms of its dominance in the habitat.
   (b) **Transformation**: Relative value (ml per unit weight)
   Reason: this transformation allows animals of different body size to be compared meaningfully without the interfering influence of actual body size.
   (c) **Transformation**: Reciprocal
   Reason: Provides a measure of rate where the data have been recorded over very different time periods (time taken for precipitation to occur). It is difficult to compare values where the time scale is different for

each recording.
   (d) **Transformation**: Rate
   Reason: Data may have been recorded over different time periods. A rate allows the production of CO_2 to be compared per unit of time over all temperatures (removes the confounding effect of different time periods as well as different temperatures).

3. Performing data transformations:
   (a) Incidence of cyanogenic clover in different regions:

Clover type	Frost free		Frost prone		Totals
	No.	%	No.	%	
Cyanogenic	124	78	26	18	150
Acyanogenic	35	22	115	82	150
Total	159	100	141	100	300

(b) Plant transpiration loss:

Time/ min	Pipette arm reading/ cm^3	Plant water loss/ $cm^3\ min^{-1}$
0	9.0	-
5	8.0	0.20
10	7.2	0.16
15	6.2	0.20
20	4.9	0.26

(c) Photosynthetic rate at different light intensities:

Light intensity/ %	Average time/ min	Reciprocal of time/ min^{-1}
100	15	0.067
50	25	0.040
25	50	0.020
11	93	0.011
6	187	0.005

(d) Frequency of size classes of eels:

Size class/ mm	Frequency	Relative frequency/ %
0-50	7	2.6
50-99	23	8.5
100-149	59	21.9
150-199	98	36.3
200-249	50	18.5
250-299	30	11.1
300-349	3	1.1
Total	**270**	**100.0**

Data Presentation (page 26)

1. The difference between the two means (labelled A) is not significant, i.e. the two means are not significantly different because the 95% CIs overlap. The mean at 4 g m^{-3} has such a large 95% CI we cannot be confident that it is significantly different from the mean at 3 g m^{-3} with the very small 95% CI.

2. Graphs and tables provide different ways of presenting information and each performs a different role. Tables summarise raw data, show any data transformations,

descriptive statistics, and results of statistical tests. They provide access to an **accurate** record of the data values (raw or calculated), which may not be easily obtained from a graph. Graphs present information in a way that makes any trends or relationships in the data apparent. Both are valuable for different reasons. **Note**: Even when you have calculated descriptive statistics for your data and tabulated these for the reader, it is a good idea to include your raw results as an appendix, or at least have them available for scrutiny.

Types of Graphs (page 27)

1. (a) A point well outside the main spread of the data.
   (b) Outliers can arise where there has been undetected experimental error (e.g. wrong reading). They can indicate sample contamination (e.g. a different species). Occasionally, they can represent real phenomena worth further study.
   (c) Where experimental error is suspected, outliers can legitimately be excluded. In other cases, it is valid to exclude extreme outliers that will bias the statistical result from an otherwise "tight" data set.

2. (a) **Line graph**. Reason: Data are continuous, and one variable is dependent on the other.
   (b) **Bar graph**. Reason: One variable is categorical (discontinuous), the other is continuous.
   (c) **Histogram**. Reason: Data are continuous. One variable is arranged in class intervals, the other is a frequency (count).

3. The class interval should reflect the degree of precision to which the data have been recorded (e.g. if weights are measured to 0.1 kg, the class interval should reflect this). An inappropriate interval may not include all data or it may be unnecessarily precise.

Drawing Scatter Plots (page 29)

1. (a) Scatter plot and (b) fitted curve below:

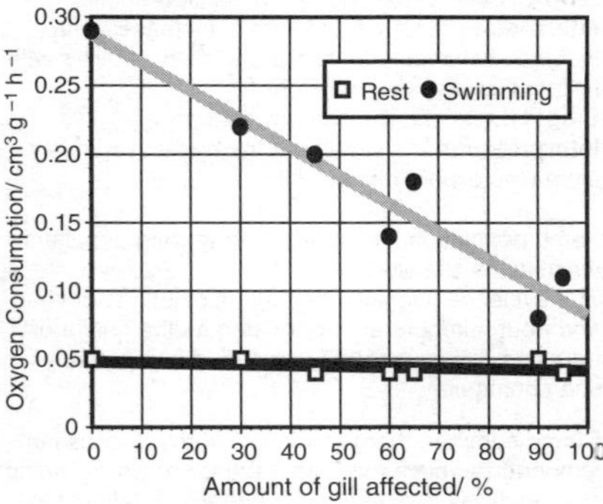

2. (a) At **rest**: No clear relationship; the line on the graph appears to have no significant slope.
   (b) **Swimming**: A negative linear relationship. The greater the proportion of affected gill, the lower the oxygen consumption.

3. The gill disease appears to have little or no effect on the oxygen uptake in resting fish.

Drawing Line Graphs (page 30)

1. (a) Line graph at the bottom of the page.
   (b) Removal of shags and nests arrowed on graph.

Interpreting Line Graphs (page 31)

1. (b) **Slope**: Negative linear relationship, with constantly falling slope.
      Interpretation: Variable Y decreases steadily with increase in variable X.
   (c) **Slope**: Constant, with slope = 0.
      Interpretation: Increase in variable X does not affect variable Y.

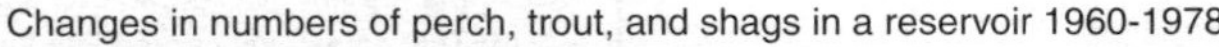

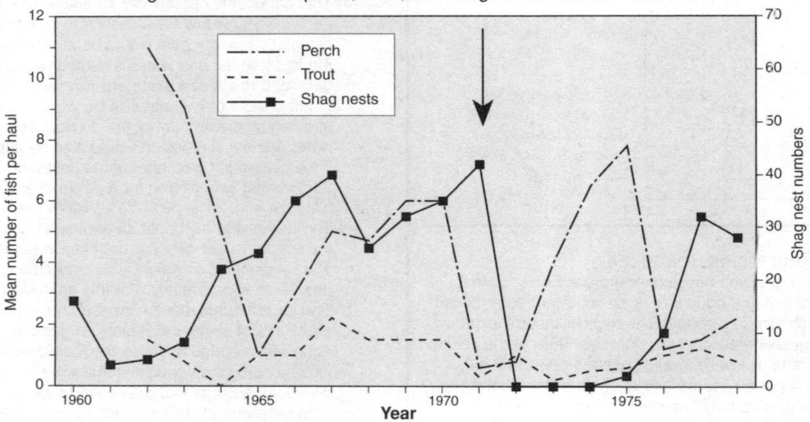

Changes in numbers of perch, trout, and shags in a reservoir 1960-1978

(d) **Slope**: Slope rises and then becomes 0.
Interpretation: Variable Y initially increases with increase in variable X, then levels out (no further increase with increase in variable X).

(e) **Slope**: Rises, peaks, and then falls (parabolic).
Interpretation: Variable Y initially increases with increase in variable X, peaks and then declines with further increase in variable X.

(f) **Slope**: Exponentially increasing slope.
Interpretation: As variable X increases, variable Y increases exponentially.

2. (a) Perch population fluctuations follow shag population fluctuations closely.
(b) The evidence suggests that the fluctuations of shag and trout numbers are not related as the height of trout fluctuations in 1967 is reached before that of shag numbers.

3. (a) During summer, the upper levels of the lake rise in temperature much more than the lower levels, which stay relatively constant, i.e. a negative relationship.
(b) The dissolved oxygen decreases as the depth increases, i.e. a negative relationship.

4. (a) During summer, both species are found in shallower water but do not extend to greater depth. In winter, both species extend to greater depths.
(b) Both species are found at greater depths in winter (June) because dissolved oxygen levels are higher to greater depth at this time (animals are not constrained by low oxygen levels).

Descriptive Statistics (page 35)

1. The modal value and associated ranked entries indicate that the variable being measured (swimmers' height) has a bimodal distribution; the data are not normally distributed. Therefore the mean and median are not accurate indicators of central tendency. **Note**: the median differs from the mean; also an indication of a skewed (non-normal) distribution.

2.

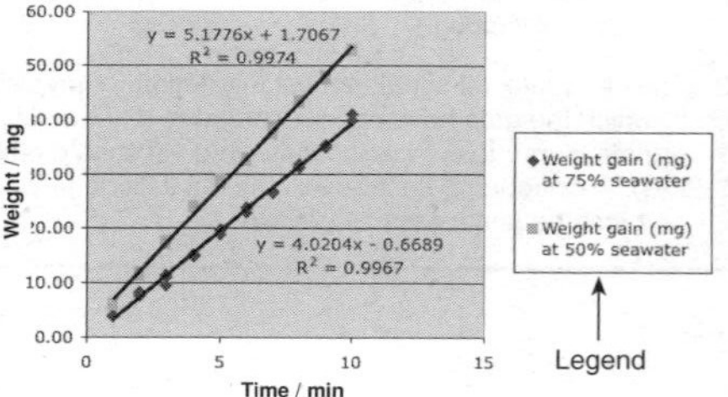

Ladybird mass (mg)	Tally	Total
6.2	\	1
6.7	I	1
7.7	\\	2
7.8	\	1
8.0	I	1
8.2	I	1
8.4	I	1
8.8	IIII	4
8.9	\	1
9.8	\	1
10.1	\	1

Median = 8th value when in rank order
$= 8.4$

Mode $= 8.8$

Mean $= 124.7 \div 15$
$= 8.3$

Note: To plot a histogram, missing weight classes would have to be included with values of O.

Linear Regression (page 39)

1. Linear regression because the plotted data fall in a straight line and body size is an important determinant of clutch size in *Daphnia*. The high R^2 value indicates the regression accounts for just over 96% of the scatter in the data. A low R^2 would have indicated that either this was not an appropriate analysis or that there was no relationship between the two variables.

2. (a) See below. This analysis is available on the TRC.

Data entry

◇	A	B	C
1	Time (min)	Weight gain (mg) at 75% seawater	Weight gain (mg) at 50% seawater
2	1	3.80	5.60
3	1	4.00	6.20
4	1	4.00	5.80
5	2	8.00	11.20
6	2	8.30	11.60
7	2	7.70	11.90
8	3	11.50	17.00
9	3	11.00	17.60
10	3	9.50	17.20
11	4	14.8	23.50
12	4	15.1	23.60
13	4	15.2	24.00
14	5	18.90	29.00
15	5	19.50	28.20
16	5	19.70	28.60
17	6	23.50	33.00
18	6	22.90	32.50
19	6	23.80	32.70
20	7	26.5	37.50
21	7	26.9	37.60
22	7	26.7	39.00
23	8	31.50	43.10
24	8	32.00	43.50
25	8	31.20	43.60
26	9	35.00	48.00
27	9	35.50	48.10
28	9	35.50	47.50
29	10	40.00	53.00
30	10	40.10	52.60
31	10	41.20	52.80
32			

Linear regression and R^2 as below. A legend added as a matter of choice.

Weight gain of crabs at two seawater dilutions

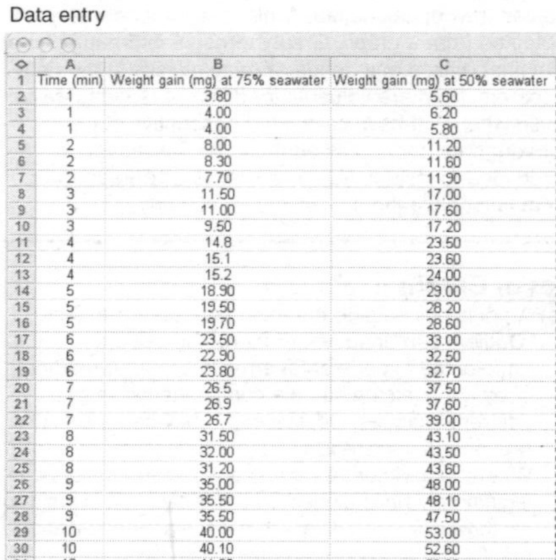

(b) Correct trend line (see above).
(c) There is a positive linear relationship for both sea water dilutions: the crabs gain weight steadily with time at both dilutions but the rate of gain is faster at the greater dilution (50% sea water).
(d) This data set is appropriate for linear regression because you would have reason to assume that cumulative weight gain would be directly related to the fact that the sea water is diluted (osmotic water gain as a result of a dilute environment). The rate of this weight gain could also be expected to be affected by the amount of the dilution. **Teacher's note**: A linear regression would also be appropriate if the net weight gain after some time interval (say 30 minutes) was plotted for a range of sea water dilutions, e.g. 25%, 50%, 75%, 100% sea water.
(e) The regression accounts for almost all (>99%) of the scatter in the data (for both sea water dilutions). The relationship of weight gain over time is described appropriately with this analysis.
(f) This experimental design involves repeated measures of single individuals. In general, this is not recommended because the individual is already influenced by the conditions that went before. In situations such as that described in this experiment, or in behavioural studies, such repeated measures

can be unavoidable, especially when resources are limited. A sample size of 3 (crabs) at each dilution is also very small. There is no control.

(g) The experimental design could be improved if recordings were made from **larger samples** (i.e. more crabs at each dilution). In the example described, there is little scatter in the data, but this would be unusual; a larger sample size is always helpful in terms of confidence of the result. There should also be a **control**, where weight gain of crabs in 100% sea water (no dilution) is measured.

The Student's *t* Test (page 42)

1. (a) The calculated *t* value is less than the critical value of $t = 2.57$. The null hypothesis cannot be rejected. (There is no difference between the control and the experimental treatments).

   (b) The new *t* value supports the alternative hypothesis at $P = 0.05$ (reject the null hypothesis and conclude that there is a difference between the control and experimental treatments). The critical value of *t* in this case is 2.23 at 10 d.f. $P = 0.05$).

2. Outliers can skew a data set, leading to mean values between data sets that are very different (even though the bulk of the data may not be very different). This could result in the rejection of the null hypothesis when it is actually true.

3. Statistical significance refers to the probability that an observed difference (or trend) will occur by chance. It is an arbitrary criterion used as the basis for accepting or rejecting the null hypothesis in an investigation. **Teacher's note**: In science the term 'significantly different' has a specific meaning. It should not be used in a casual manner when no statistical test has been performed.

Student's *t* Test Exercise (page 43)

1. (a) Completed table below

| X (counts) | | $x - \bar{x}$ | | $(x - \bar{x})^2$ | |
Popn A	Popn B	Popn A	Popn B	Popn A	Popn B
465	310	9.3	-10.6	86.5	112.4
475	310	19.3	-10.6	372.5	112.4
415	290	-40.7	-30.6	1656.49	936.36
480	355	24.3	34.4	590.49	1183.36
436	350	-19.7	29.4	388.09	864.36
435	335	-20.7	14.4	428.49	207.36
445	295	-10.7	-25.6	114.49	655.36
460	315	4.3	-5.6	18.49	31.36
471	316	15.3	-4.6	234.09	21.16
475	330	19.3	9.4	372.49	88.36
$n_A = 10$ $n_B = 10$				$\Sigma(x - \bar{x})^2$	$\Sigma(x - \bar{x})^2$
The number of samples in each data set				4262.1	4212.4

   (b) Variance of population A: 473.57
   Variance of population B: 468.04
   (c) Difference between population means: 135.1
   (d) $t = 13.92$
   (e) Degrees of freedom: 18
   (f) $P = 0.05$, *t* (critical value) = 2.101
   (g) **Decision**: We can reject the null hypothesis. The difference between the population means is significantly different at $P = 0.05$. **Note** at $P = 0.001$,

the critical *t* value is 3.922, so we can also reject the null hypothesis at $P = 0.001$.

2. (a) Spreadsheet to be completed by the student. Calculated values as follows:
   Mean: A = 455.7, B = 320.6
   Difference between means: 135.1
   Sum of squares: A = 4262.1, B = 4212.4
   Variance: A = 473.567, B = 468.044
   $t = 13.92$

   (b) New *t* value: 0.76. Decision: We cannot reject H_0 ($P = 0.05$). There is no difference between population means.

Comparing More Than Two Groups (page 45)

1. Number of treatments: 5

2. Independent variable: Pasture type

3. Dependent variable: Survival after 14 days

4. A column chart is a suitable graph to plot this data set; the independent variable is categorical and the dependent variable is continuous.

5. The 95% CI gives an indication of the reliability of the sample mean as an estimate of the true population mean. The 95% CI can be used to determine the 95% confidence limits (mean ± 95% CI). These tell you that 95% of the time, the true population will lie within these limits. (This is equivalent of a $P = 0.05$).

Analysis of Variance (page 47)

1. Although ANOVA can not tell us which one (or more) of the groups is significantly different from the others, we can see from the plot with 95% CI that survival of weevils on white clover was significantly better than on the other pasture types (the bars of the 95% CI for white clover are well separated from those of the other pasture types). Note: The 95% CI also tells us that there is no significant difference between the other pasture types with respect to weevil survival.

Writing Your Report (page 48)

Report checklist to be completed by the student.

Writing the Methods (page 49)

1. (a)-(d) Mistakes (any four):
   - The transect line should be marked out at intervals, rather than having to 'step out' the distances.
   - Transect length is not given.
   - The 100 paces would not be consistent intervals as different people were stepping them out.
   - The biggest plant under the string should not be the one that is recorded; the one directly under the string at a predetermined point (e.g. every 1 m) should be recorded.
   - Bare ground and unidentifiable plants must be recorded on a tally chart of plants present.
   - The repeat should be carried out parallel to the first transect line and not just at another 'similar' area.
   - The transect line should have been placed in the sun and the shade separately, not half in the sun and half in the shade. It would not be easy to tell where these boundaries lay or if they were static.

- There is no quantitative measure (or even qualitative assessment) of sun and shade. The time of year during which the sampling occurred is not stated, yet is likely to be important.
- No tally sheet is given in which to record the data and no formula is given for the calculation of percentage abundance.

Writing Your Results (page 50)

1. Referring to tables and figures in the text clearly indicates which data you are referring to in your synopsis of the results and gives the reader access to these data so that they can assess your interpretation.

2. Tables summarise data and provide a record of the data values, which may not be easily obtained from a graph. Figures (graphs) present information in a way that makes trends or relationships in the data apparent. Such trends may not be evident from the tabulated data. Both formats are valuable for different reasons.

Writing Your Discussion (page 51)

1. Discussion of weaknesses in your study shows that you have considered these and acknowledged them and the effect that they may have had on the outcome of your investigation. It also provides the opportunity for those repeating the investigation (including yourself) to improve on aspects of the design.

2. A **critical evaluation** shows that you have examined your results carefully in light of the question(s) you asked and the predictions you made. Objective evaluation enables you to provide reasonable explanations for any unexpected or conflicting results and identify ways in which to improve your study design in future investigations.

3. The conclusion allows you to make a clear statement about your findings, i.e. whether for not the results support your hypothesis. If your results and discussion have been convincing, the reader should be in agreement with the conclusion you make.

Citing and Listing References (page 52)

1. A bibliography lists all sources of information whereas a reference list includes only those sources that are cited in the text. Usually a bibliography is used to compile the final reference list, which appears in the report.

2. Internet articles can be updated as new information becomes available and the original account is revised. It is important that this is noted because people using that source in the future may find information that was unavailable to the author making the original citation.

3. Reference list as follows:
   Ball, P. (1996): Living factories. New Scientist, 2015, 28-31
   Campbell, N. (1993): Biology. Benjamin/Cummings. Ca.
   Cooper, G. (1997): The cell: a molecular approach. ASM Press, Washington DC. pp. 75-85
   Moore, P. (1996): Fuelled for Life. New Scientist, 2012, 1-4
   O'Hare, L. & O'Hare, K. (1996): Food biotechnology. Biological Sciences Review, 8(3), 25.
   Roberts, I. & Taylor, S. (1996): Development of a procedure for purification of a recombinant therapeutic protein. Australasian Biotechnology, 6(2), 93-99.

KEY TERMS Mix and Match (page 54)

95% CI (S), Anova (K), Assumption (G), Bibliography (J), Continuous data (Q), Control (F), Discontinuous data (B), Experiment (L), Graph (H), Hypothesis (D), Mean (A), Median (E), Mode (X), Null hypothesis (N), Qualitative data (V), Quantitative data (P), Range (I), Report (W), Sample (M), Scientific method (O), Standard deviation (C), Standard error (Y), Table (R), Trend (U), Variable (T).

ATP in Metabolism (page 57)

1. In the presence of the enzyme ATPase, ATP is hydrolysed to produce ADP plus a free phosphate, releasing energy in the process.

2. Glucose

3. Cellular respiration, strictly oxidative phosphorylation.

4. Solar energy

5. Food (gaining nutrient from plants and other animals)

6. Like a rechargeable battery, the ADP/ATP system toggles between a high energy state and a low energy. The addition of a phosphate to ADP recharges the molecule so that it can be used for cellular work.

7. **PHOTOSYNTHESIS**
   Starting materials: Carbon dioxide, water (as a source of hydrogens), in the presence of light and chlorophyll.
   Waste products: Oxygen, water.
   Role of hydrogen carriers: NADP: Carries hydrogen between light dependent and light independent phases (where the hydrogen is incorporated into sugars).
   Role of ATP: Produced in the light dependent phase and used in the light independent phase to make sugars from carbon dioxide and hydrogen.
   Overall biological role: Uses light energy to fix carbon into organic molecules which become part of the energy available in food chains.
   CELLULAR RESPIRATION
   Starting materials: Organic molecules (ultimately glucose), oxygen.
   Waste products: Carbon dioxide, water.
   Role of hydrogen carriers: NAD: Carries hydrogens to the electron transport chain, where their transfer between carriers is coupled to ATP production.
   Role of ATP: A small amount of ATP is used initially to produce pyruvate from glucose. Produced in glycolysis, Krebs cycle, and the ETS.
   Overall biological role: The process by which organisms break down energy rich molecules to release energy in a usable form (ATP).

Plants as Producers (page 59)

1. (a) water + carbon dioxide (in the presence of light and chlorophyll) $\rightarrow$ glucose + oxygen + water
   (b) $12H_2O + 6CO_2 \rightarrow C_6H_{12}O_6 + 6O_2 + 6H_2O$

2. Plants provide oxygen to the atmosphere, remove CO_2 and help prevent its build up, and provide the food base for the majority of the Earth's ecosystems.

3. Algae are producers and need to remain in the euphotic zone; the zone where there is enough light penetration for photosynthesis to occur (0-30 m).

4. As deforestation progresses, oxygen becomes in shorter supply, carbon dioxide levels in the atmosphere build up with fewer plants around to act as carbon sinks, and the climate warms progressively.

Summary of Photosynthesis (page 60)

1. (a) Grana: Membranes that contain chlorophyll and are the site of the light dependent phase of photosynthesis. The biochemical process involves energy capture via photosystems I and II.
   (b) Stroma: The liquid interior of the chloroplast in which the light independent phase takes place. The biochemical process involves carbon fixation via the Calvin cycle.

2. (a) Carbon dioxide: Comes from the air (via stomata) and provides carbon and oxygen as raw materials for the production of glucose. Some oxygen molecules contribute to the production of H_2O
   (b) Oxygen: Comes from CO_2 gas (via stomata) and H_2O (via the roots and vascular system). The oxygen from the CO_2 is incorporated into glucose and H_2O. The oxygen from the water is given off as free oxygen (a waste product).
   (c) Hydrogen: Comes from water (via the roots and vascular system) obtained from the soil. This hydrogen is incorporated into glucose and H_2O.
   Note: The clarify: isotope studies show that the carbon and oxygen in the carbohydrate comes from CO_2, while the free oxygen comes from H_2O.

3. Scientists use isotopes to tag molecules taken up by plants and analyse the products produced as a result of photosynthesis. For example, the use of oxygen isotopes can identify the origin of the free oxygen released in photosynthesis:
   $$CO_2 + 2H_2{}^{18}O \rightarrow [CH_2O] + {}^{18}O_2 + H_2O$$

4. Glucose is used as the fuel for cellular respiration, or used to construct cellulose, starch, or disaccharide molecules (e.g. fructose). Oxygen is required for aerobic respiration and water is recycled and even reused for photosynthesis.

Chloroplasts (page 61)

1. (a) Stroma (d) Granum
   (b) Lamallae (e) Thylakoid
   (c) Outer membrane (f) Inner membrane

2. (a) Chlorophyll is found in the thylakoid membrane.
   (b) Chlorophyll is a membrane-bound pigment found in and around the photosystems that embedded in the membranes. Light capture by chlorophyll is linked to electron transport in the light dependent reactions.

3. The internal membranes provide a large surface area for binding chlorophyll molecules and capturing light. Membranes are stacked in such a way that they do not shade each other.

4. Chlorophyll absorbs blue and red light but reflects green light, so leaves look green to the human eye.

Pigments and Light Absorption (page 62)

1. The **absorption spectrum** of a pigment is that wavelength of the light spectrum absorbed by a pigment, e.g. chlorophyll absorbs red and blue light and appears green. Represented graphically, the absorption spectrum shows the relative amounts of light absorbed at different wavelengths.

2. Accessory pigments absorb light wavelengths that chlorophyll *a* cannot absorb, and they pass their energy on to chlorophyll *a*. This broadens the action spectrum over which chlorophyll *a* can fuel photosynthesis.

Light Dependent Reactions (page 63)

1. **NADP**: Carries H_2 from the light dependent phase to the light independent reactions.

2. **Chlorophyll**: These pigment molecules trap light energy and produce high energy electrons. These are used to make ATP and NADPH. The chlorophyll molecules also split water, releasing H^+ for use in the light independent reactions and liberating free O_2.

3. Light dependent (D) phase takes place in the grana (thylakoid membranes) of the chloroplast and requires light energy to proceed. The light dependent phase generates ATP and reducing power in the form of NADPH. The electrons and hydrogen ions come from the splitting of water.

4. The ATP synthesis is coupled to electron transport. When the light strikes the chlorophyll molecules, high energy electrons are released by the chlorophyll molecules. The energy lost when the electrons are passed through a series of electron carriers is used to bond a phosphate to ADP to make ATP.
   Note: ATP is generated (in photosynthesis and cellular respiration) by **chemiosmosis**. As the electron carriers pick up the electrons, protons (H^+) pass into the space inside the thylakoid, creating a high concentration of protons there. The protons return across the thylakoid membrane down a concentration gradient via the enzyme complex, ATP synthetase that synthesises the ATP (also called ATP synthase or ATPase).

5. (a) **Non-cyclic (photo)phosphorylation**: Generation of ATP using light energy during photosynthesis. The electrons lost during this process are replaced by the splitting of water.
   (b) The term non-cyclic **photo**phosphorylation is also (commonly) used because it indicates that the energy for the phosphorylation is coming from light.

6. (a) In **cyclic photophosphorylation**, the electrons lost from photosystem II are replaced by those from photosystem I rather than from the splitting of water. ATP is generated in this process, but not NADPH. **Note**: In the cell, both cyclic and non-cyclic photophosphorylation operate to different degrees in order to keep the production of NADPH and ATP balanced.
   (b) The non-cyclic path produces ATP and NADH in roughly equal quantities but the Calvin cycle uses more ATP than NADPH. The cyclic pathway of electron flow makes up the difference.

7. It shows that a complex reaction pathway is made of less complex pathways that can operate independently.

These simple pathways can then be linked through common intermediates, generating complex pathways.

Light Independent Reactions (page 65)

1. (a) 6   (b) 6   (c) 12
   (d) 12   (e) 12   (f) 6
   (g) 2   (h) 1

2. RuBisCo catalyses the reaction that splits CO_2 and joins it with ribulose 1,5-bisphosphate. It fixes carbon from the atmosphere.

3. Glyceraldehyde - 3 - phosphate (GALP)

4. $6CO_2 + 18ATP + 12\ NADPH + 12H^+$
   $\rightarrow 1\ glucose + 18ADP + 18P_i + 12\ NADP^+ + 6H_2O$

5. The Calvin cycle will cease in the dark in most plants because the light dependent reactions stop, therefore no NADPH or ATP is produced. At night, stomata also close, reducing levels ofCO_2 (there will still be some CO2 in the leaf as a waste product of respiration).

Components of an Ecosystem (page 66)

1. A **community** is a naturally occurring group of organisms living together as an ecological entity. The community is the biological part of the ecosystem. The **ecosystem** includes all of the organisms (the community) and their physical environment.

2. The **biotic factors** are the influences that result from the activities of living organisms in the community whereas the **abiotic** (physical) **factors** comprise the non-living part of the community, e.g. climate.

3. (a) Population   (c) The community
   (b) Ecosystem   (d) Physical factor

Energy Inputs and Outputs (page 67)

1. **Producers** convert energy received from an inorganic source (usually sunlight) into a form that is accessible to consumer levels. **Consumers** depend on the energy stored in the chemical bonds of biological molecules (the fats, proteins, and carbohydrates of plant and animal tissues). They too transfer energy to other levels, but energy is lost with each transfer.

2. In a grazing food web, energy moves from producers (plants) to primary consumers (herbivores) and then to secondary consumers (carnivores). This chain of energy transfer can continue several times, but eventually ends. All these consumer groups provide energy to decomposer levels. In a detrital food web, producers provide energy as dead plant material, and the primary consumers are decomposer microbes such as bacteria and fungi. Energy flows back and forth between decomposers and detritivores but herbivores and carnivores do not feature.

3. Detritivores consume (ingest) detritus and, in doing so, speed up decomposition by increasing the surface area available to decomposer bacteria. Decomposers (bacteria and fungi) also use detritus as an energy source but digestion is extracellular (enzymes are secreted in fungi or bound to the cell surface in bacteria). These enzymes break down the detritus into constituent molecules for absorption so the breakdown is more complete than is the case with detritivores.

Food Chains (page 68)

1. (a) The sun.
   (b)-(d): See diagram at the bottom of the page.

2. (a) Each successive trophic level has less energy.
   (b) Energy is lost by respiration as it is passed from one trophic level to the next.

3. A **food chain** comprises a sequence of organisms, each of which is a source of food for the next. Food chains in ecosystems are organised according to **trophic levels**; the feeding levels that energy passes through as it proceeds through the food chain. Organisms are assigned a category according to the trophic level they occupy. Producers form the first trophic level, 1st order consumers (primary consumers) eat producers (i.e they are herbivores), 2nd order consumers (secondary consumers) eat herbivores (i.e. they are carnivores) etc. Organisms may occupy more than one trophic level depending on diet. Detritivores and decomposers obtain energy from all other trophic levels and are therefore not assigned a trophic level.

4. The kingfisher occupies different trophic levels at different times (depending on the prey of choice).

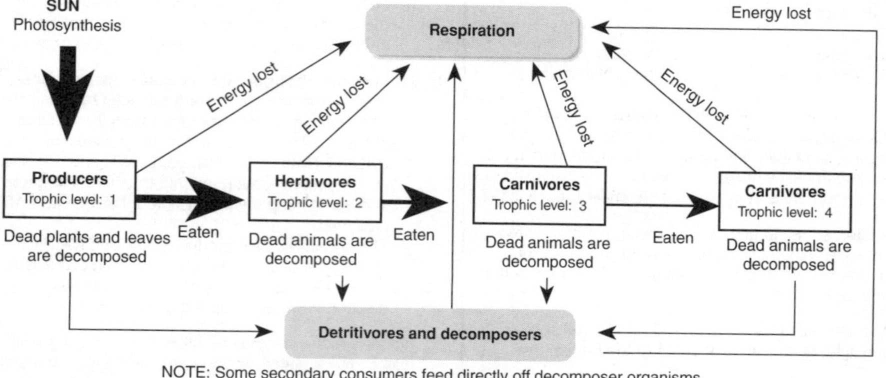

Primary Productivity (page 69)

1. Primary productivity affected by ((a)-(c) any three of):
   - The mass of plant material (specifically, the actively growing plant mass). This is the plant material that produces new biomass per unit time.
   - Nutrients/water availability: Limiting amounts of these will limit primary productivity.
   - Leaf area index (see question 3).
   - Light: The primary productivity of light limited systems will be limited and determined by the maximum photosynthetic rates that can be achieved in the light available.
   - Temperature: Low temperature limits the primary productivity of systems because growth rates are slowed by low temperatures.

2. **Production** refers to the total energy (biomass) fixed through photosynthesis, whereas **productivity** refers to the rate at which this occurs, i.e. the rate of production.

3. LAI is an indirect measure of the capacity of the plant to intercept light, photosynthesise, and produce new plant biomass. High LAIs may be typical of high productivity, although this is not necessarily the case. Large trees may support a large biomass with relatively low rates of production of new biomass. Conversely, grasslands may have high rates of production despite cropping because they are not supporting a large, static biomass. The concept is more applicable in agricultural situations where there can be marked **reductions** in the (typical) LAI for a crop, e.g. through crop damage due to pests, and this reduces the primary productivity. **Note**: The LAI is the measure of the leaf area of a plant exposed to incoming light, expressed in relation to the ground surface area beneath the plant.

4. Graph below. Both data sets have been plotted here.

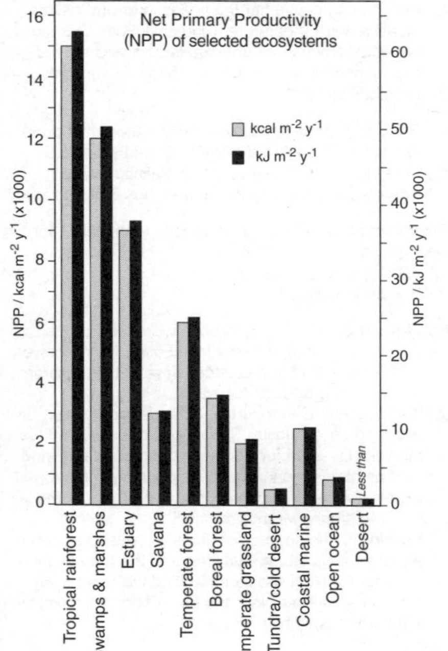

5. (a) The key factors limiting rates of primary production in **terrestrial** ecosystems are temperature and moisture; the productivity of tundra ecosystems is limited by low temperatures, while that of desert ecosystems is limited by moisture availability. Tropical rainforest ecosystems do not have these same limitations.
   (b) In aquatic systems, light and nutrient availability limit rates of production. The NPP of open ocean is low relative to coastal systems because of the low levels of nutrients. Nitrogen and phosphorus, in particular, are very low in the open ocean but higher in coastal systems which receive inputs from the land. **Note**: Although light may be limiting to productivity in the open ocean, tropical waters are less productive than one would predict from the higher light intensities there; low nutrient availability is the critical factor in this case.

6. (a) The amount of water contained in plants is variable and unrelated to their energy content, hence the dry weight (not wet weight) is required.
   (b) **Standing crop**: The mass of all the plant material (including that underground) from a measured area.
   (c) The procedure outlined only gives an estimate of NPP because you cannot easily determine the biomass not collected due to consumption or death.
   (d) To express standing crop in kJ m⁻² you would need to know the energy content of the plant material (per known mass unit measured, e.g. kg). **Note**: A calorimeter could be used to do this.
   (e) To calculate GPP you would need to know the energy lost in respiration, e.g. by measuring the CO_2 lost at night.

7. (a) High rates of production are achieved using inputs of fertilisers, intensive pest control, and irrigation. Large inputs of energy (energy subsidies) are required for this; fossil fuels are used to drive machinery, produce pesticides and fertilisers, and breed high yielding plant varieties.
   (b) These energy inputs are, in the long term, not sustainable; effort must be placed in using sustainable farm practices (intercropping, crop rotation, mulching, green cropping etc.) to reduce energy cost of high production.

Energy Flow in an Ecosystem (page 71)

1. (a) 14 000   (b) 180   (c) 35   (d) 100

2. Solar energy

3. A. Photosynthesis
   B. Eating/feeding/ingestion
   C. Respiration
   D. Export (lost from this ecosystem to another)
   E. Decomposers and detritivores feeding on other decomposers and detritivores
   F. Radiation of heat to the atmosphere
   G. Excretion/egestion/death

4. (a) 1 700 000 ÷ 7 000 000 x 100 = 24.28%
   (b) It is reflected. Plants appear green because those wavelengths are not absorbed. Reflected light falls on other objects as well as back into space.

5. (a) 87 400 ÷ 1 700 000 x 100 = 5.14%
   (b) 1 700 000 - 87 400 = 1 612 600 (94.86%)
   (c) Most of the energy absorbed by the producers is

not used in photosynthesis. This excess energy which is not fixed is lost as heat (although the heat loss component **before** the producer level is not usually shown on energy flow diagrams). **Note**: Some of the light energy absorbed through accessory pigments widens the spectrum that can drive photosynthesis. However, much of accessory pigment activity is associated with photoprotection; they absorb and dissipate excess light energy that would otherwise damage chlorophyll.

6. (a) 78 835 kJ
    (b) 78 835 ÷ 1 700 000 x 100 = 4.64%

7. (a) Decomposers and detritivores
    (b) Transport by wind or water to another ecosystem.

8. (a) Energy remains locked up in the detrital material and is not released.
    (b) Geological reservoir:

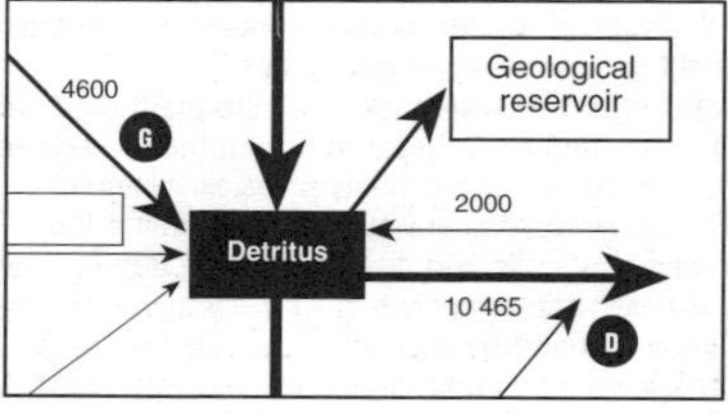

(c) **Oil** (petroleum) and **natural gas**, formed from the remains of marine plankton. **Coal** and **peat** are both of plant origin; peat is partly decomposed, and coal is fossilised.

9. (a) 87 400 → 14 000: 14 000 ÷ 87 400 x 100 = 16%
    (b) 14 000 → 1600: 1600 ÷ 14 000 x 100 = 11.4%
    (c) 1600 → 90: 90 ÷ 1600 x 100 = 5.6%

The Carbon Cycle (page 73)

1. Arrows can be added for the points (a)-(d) as follows:
    (a) Dissolving of limestone by acid rain: Arrow from the limestone layer to atmospheric CO_2.
    (b) Release of carbon from the marine food chain: Arrows (labelled **respiration**) from marine organisms (shark, algae, fish) to atmospheric CO_2.
    (c) Mining and burning of coal: Arrow from the coal seam to atmospheric CO_2.
    (d) Burning of plant material: Arrow (labelled **combustion**) from the trees and/or grassland to atmospheric CO_2.

2. (a) Coal: Plant material trapped under sediment in swampy conditions millions of years ago.
    (b) Oil: Marine plankton rapidly buried in sediment mya.
    (c) Limestone (also chalk = fine limestone): Shells of molluscs, skeletons of coral and other marine organisms with skeletons of calcium carbonate piled upon seabeds and compressed.

3. (a) Photosynthesis   (b) Respiration

4. Carbon would eventually be locked up in the bodies (remains) of dead organisms. Dead matter would not rot. Possible gradual loss of CO_2 from the atmosphere.

5. (a) **Respiration** (stepwise oxidation of glucose)
    (b) **Combustion** (rapid oxidation of organic substances

accompanied by heat).

6. (a) Humans deplete fossil fuel reserves through mining (they provide a source of readily available energy).
    (b) The burning of fossil fuels increases the amount of carbon dioxide in the atmosphere, contributing to the rise in global temperatures. Burning also increases levels of air pollution.
    (c) Minimising fossil fuels use through the use of alternative, environmentally clean sources of energy (solar energy, wind energy). Making sure that when fossil fuels are burnt, that combustion is as clean (complete) as possible, to minimise pollution. Using biofuels to replace fossil fuels as an energy source where possible.

7. (a) Deforestation (either by cutting or burning) increases atmospheric CO_2 levels (the CO_2 stored in the trees is released upon burning or cutting).
    (b) Planting more trees (reforestation) will decrease atmospheric CO_2 (the trees utilise the CO_2 during photosynthesis and store it as carbohydrate).

Physical Factors and Gradients (page 75)

1. Environmental gradients:
    (a) Salinity: Increases from LWM to HWM
    (b) Temperature: Increases from LWM to HWM
    (c) Dissolved oxygen: Decreases from LWM to HWM
    (d) Exposure: Increases from LWM to HWM

2. (a) **Mechanical force of wave action**: Point B will receive the full force of waves moving inshore, Point A will receive only milder backwash, Point C will experience some surge but no direct wave impacts. **Surface temperature**: Points A and B will experience greater variations in rock temperature depending on whether the tide is in or out, day or night, water temperature, wind chill. Point C is more protected from some of these factors and will not experience the warming effect of direct sunlight.
    (b) Microclimate.

3. (a) Rock pools may have low salinity if there has been rain falling into them directly or through runoff.
    (b) Rock pools may have very high salinity due to evaporation after long exposure times without rain.

4. Environmental gradients from canopy to the leaf litter:
    (a) Light intensity: decreases
    (b) Wind speed: decreases
    (c) Humidity: increases.

5. (a) **Light intensity**: Foliage above will shade plants below, with a cumulative effect. Forest floor receives light reflected off leaf surfaces several times, some even passing through leaves.
    (b) **Wind speed**: Canopy trees act as a wind-break, reducing wind velocity. Subcanopy trees will reduce the velocity even further; near the ground the wind may be almost nonexistent. An opening in the forest canopy (a clearing) can expose the interior of the forest to higher wind velocities.
    (c) **Humidity**: The sources of humidity (water vapour) are the soil moisture, leaf litter and the transpiration from plants. Near the canopy, wind will carry away moisture-laden air. Near the forest floor, there is little wind, and humidity is high.

6. Light quality: the colour of the light will change near

the forest floor. White light falling on the canopy will be absorbed by the leaves. Reflected light in the green wavelength bounces off the leaves and passes to lower foliage and the forest floor.

7. (a) Advantage: one of: Reduced wind speed reduces transpirational water loss. Higher humidity is advantageous for plants sensitive to water loss.
   (b) Disadvantage: There is a marked reduction in the quantity and quality of light available for photosynthesis.

8. Environmental gradients:
   (a) Water temperature: Decreases gradually until below the zone of mixing when there is a sharp drop.
   (b) Dissolved oxygen: At a uniform concentration until below the zone of mixing when there is a sharp drop, with very little oxygen at the bottom.
   (c) Light penetration: Decreases at an exponential rate (most light is absorbed near the surface).

9. (a) Prevents mixing of the O_2-rich surface water with the deeper O_2-deficient water (a thermal barrier). water below the thermocline can become anoxic.
   (b) Organisms living below the thermocline use up much of the available oxygen. Decomposition also uses up oxygen.

10. (a) **Heavy rainfall** or inflow of floodwater from a major river channel nearby dilutes salts in the lake.
    (b) **Evaporation** from the lake concentrates salts.

11. Physical gradients govern what organisms will be found and where in a particular area, as determined by the tolerance levels of individual organisms.

Habitat (page 78)

1. An organism will occupy habitat according to its range of tolerance for a particular suite of conditions (temperature, vegetation and cover, pH, conductivity). Organisms will tend to occupy those regions where all or most of their requirements are met and will avoid those regions where they are not. Sometimes, a single factor, e.g. pH for an aquatic organisms, will limit occupation of an otherwise suitable habitat.

2. (a) Most of a species population is found in the optimum range because this is the zone where conditions for that species are best; most of the population will select that zone.
   (b) The greatest constraint on an organism's growth within its optimum range would be intraspecific competition (or competition with different species with similar niche requirements).

3. In a marginal niche, any of the following might apply:
   - Physicochemical conditions (e.g. temperature, current speed, pH, conductivity) might be sub-optimal and create stress (therefore greater vulnerability to disease).
   - Food might be more scarce or of lower quality.
   - Mates might be harder to find.
   - The area might be more exposed to predators.
   - Resting, sleeping, or nesting places might be harder to find and/or less suitable in terms of shelter or safety.
   - Competition from other better-adapted species might be more intense.

Features of Populations (page 79)

1. (a) One of the following:
   Population growth rate: If this increases (or decreases) from one time interval to the next, it indicates that the population is probably also increasing (or decreasing). **Note**: The **intrinsic** rate of population increase (r_{max}) should be distinguished from population growth rates that account for the increasing number of individuals in the population (rN). The intrinsic population growth rate is a characteristic value for each species but rN can increase rapidly as more and more individuals add to the population increase.
   Total abundance: If this increases (or decreases) from one time interval to the next, it indicates that the population is also increasing (or decreasing).
   Mortality: If this is increasing from one time interval to the next, it indicates that the population may be decreasing (you must also account for other sources of population change).
   Birth rate & population fertility: If these increase from one time interval to the next, they indicate that the population may be increasing (you must also account for other sources of population change).
   Age structure: A population dominated by young individuals is usually increasing. A population dominated by old (especially post-reproductive) individuals is usually decreasing.
   (b) One of the following:
   Distribution: A very clumped distribution may indicate that only some parts of the environment are suitable for supporting individuals.
   Population growth and birth rates: If these are low or declining it may indicate an inability of the environment to support the population density.
   Mortality: If this are very high or increasing it *may* indicate an inability of the environment to support the present population density.

2. (a) **Measurable attributes**: Density, distribution, total abundance, sex ratios, migration (sometimes difficult). In some cases, depending on the organism, also age structure and population fertility.
   (b) **Calculated attributes**: Population growth rate, birth rate (natality) and death rate (mortality).

3. (a) Population sampling of an endangered species allows us to determine (any of): If population is growing and how fast; the population's age and sex structure (i.e. is it dominated by young or old, non-reproductive, individuals); population abundance, density, and distribution in different areas (habitat preference and suitability); sources of mortality (predation, disease, starvation etc.); population fertility. This information allows informed decisions to be made about the population's management.
   (b) Population sampling of a managed fish species allows us to determine the population growth rate. This is critical to establishing the level of fishing that can be supported by the population (the sustainable harvest) without irreversible population decline. The growth rate is calculated taking into account population abundance, and birth and death rates. Sustainable harvest can be built into the equation as one of the (controllable) sources of mortality.

Density and Distribution (page 80)

1. (a) Resources such as food and shelter are not usually spread through the environment in an even manner. Organisms will clump around these resources.
   (b) Some organisms group together for protection from the physical environment or from predators, or for the purposes of reproduction. Clumped distribution may also result from the dispersal method, especially in plants where reproduction may occur by seeds (widely dispersed) or vegetative methods.

2. Territorial behaviour whereby the establishment and demarcation of an area distributes individuals uniformly.

3. Resources are limited but distributed uniformly.

4. (a) **Clumped**: Many marine gastropods, colonial birds (seasonally), many mammals that exhibit grouping/ herd behaviour, schooling fish, colonial insects, many other invertebrates such as coral, some plants with limited dispersal.
   (b) **Random**: Weed plants with effective dispersal method, shellfish on sand or mud substrate.
   (c) **Uniform**: Territorial organisms, monoculture plantings (e.g. crops, timber plantations).

Resources and Distribution (page 81)

1. Availability of resources in habitat. Regions with poor food sources, and few watering holes or sleeping trees, by necessity must be larger.

2. • Promotes division of labour within the group, maximising the efficiency with which the group can search for food and defend itself. • Reduces aggression arising from resource competition: resources are allocated according to hierarchical status and more energy can be used to procure food.

3. The troop will defend the core area aggressively because it contains the most valuable resources (to which exclusive access is desirable). In terms of energy expenditure, the effort expended in defence is worth it.

4. (a) 8 home ranges
   (b) Point A: 2   Point B: 1
   (c) There is a large degree of overlap in the **home ranges** of neighbouring troops. **Core areas** do not appear to overlap with neighbouring troops and are separated some distance from each other.

5. Generally:
   – **Many passerine (perching) birds**: Territory provides an area for resource acquisition, courtship, mating, and rearing chicks during the breeding season.
   – **Many marine mammals** show female groups with male territoriality. The territoriality in this case involves an area with resources (a safe site for breeding) desired by the females. By controlling these resources, the males gain access to mates. Examples: fur seals, sea lions, elephant seals.
   – **Mammalian canids and big cats**: Many mammalian predators, e.g. wolves, hyenas, lions, live in cooperative social groups that defend a permanent (hyena) or seasonal (wolves) territory, often with a centrally located den or breeding site, against other such groups (and against other directly competing species). The territory provides access to game without interference from other groups and provides some degree of protection

for the young, which can be protected more easily within the defended area.

Population Regulation (page 82)

1. (a) Density dependent factor: Predation (e.g. by ladybird beetles), competition with other aphids for position on the best part of the plant to feed.
   (b) Density independent factor: Temperature (drop in temperature in autumn months in cooler climates causes the population to crash).

2. When population density is low, individuals are well spaced apart. This can reduce stress between individuals (improving the resistance to diseases) as well as making the transmission of the pathogen more difficult. Crowded populations are more susceptible to epidemics of infectious disease.

3. **Density dependent** factors, such as disease, parasitic infestation, competition, and predation have an increasing effect on population growth as the density of the population increases; their effects are exacerbated at high population densities because they are driven in part by the number of organisms present. **Density independent** factors, such as flood, fire, and drought have a controlling effect on population size and growth that is independent of the population density. The severity of the impact on the population is not correlated with the density of the population.

Population Growth (page 83)

1. (a) Mortality: Number of individuals dying per unit time (death rate).
   (b) Natality: Number of individuals born per unit time (birth rate).
   (c) Net migration rate: Net change in population size per unit time due to immigration and emigration.

2. Population growth will be constrained to a level that can be sustained by the factor that is most limiting.

3. (b) A declining population: $B + I < D + E$
   (c) An increasing population: $B + I > D + E$

4. (a) Birth rate = 14 births ÷ 100 total number of individuals x 100 % = 14% per year
   (b) Net migration rate = 2% per year
   (c) Death rate = 20% per year
   (d) Rate of population change: birth rate – death rate + net migration rate = $14 - 20 + 2 = -4\%$ per year
   (e) The population is **declining**.

5. Limiting factors to human population growth: availability of food or land (to live or grow food), prevalence of disease and ability of public health system to prevent and treat it.

Niche Differentiation (page 84)

1. (a) Different species may exploit different microhabitats within the ecosystem (e.g. tree trunks, leaf litter, lower or upper canopy).
   (b) Different species may exploit the same resources but at different times of the day or year.

2. Competition is avoided by the birds exploiting slightly different parts of the eucalypt habitat, with the birds

feeding in different ways and/or on different food types.

3. Each damselfish exploits different parts of the reef crest and slopes as well as feeding on different food supplies and/or at different times of the day.

Intraspecific Competition (page 85)

1. (a) **Individual growth rate**: Intraspecific competition may reduce individual growth rate when there are insufficient resources for all individuals (Examples: tadpoles, *Daphnia*, many mammals with large litters). **Note**: Individuals compete for limited resources and growth is limited in those individuals that do not get access to sufficient food.

   (b) **Population growth rate**: Intraspecific competition reduces population growth rate. Examples as above. **Note**: Competition intensifies with increasing population size and, at carrying capacity, the rate of population increase slows to zero.

   (c) **Final population size**: Intraspecific competition will limit population size to a level that can be supported by the carrying capacity of the environment. **Note**: In territorial species, this will be determined by the number of suitable territories that can be supported.

2. (a) They reduce their individual growth rate and take longer to reach the size for metamorphosis.
   (b) Density dependent.
   (c) The results of this tank experiment are unlikely to represent a real situation in that the tank tadpoles are not subject to normal sources of mortality and there is no indication of long term survivability (of the growth retarded tadpoles). **Note**: At high densities, many tadpoles would fail to reproduce and this would naturally limit population growth (and size) in the longer term.

3. Reduce intensity of intraspecific competition by:
   (a) Establishing hierarchies within a social group to give orderly access to resources.
   (b) Establishing territories to defend the resource within a specified area.

4. (a) Carrying capacity might decline as a result of unfavourable climatic events (drought, flood etc.) or loss of a major primary producer (plant species).
   (b) Final population size would be smaller (relative to what it was when carrying capacity was higher).

5. **Territoriality** is a common consequence of intraspecific competition in mammals and birds. In any habitat, resources are limited and only those with sufficient resources will be able to breed. This is especially the case with mammals and birds, where the costs of reproduction to the individual are high relative to some other taxa. Even though energy must be used in establishing and maintaining a territory, territoriality is energy efficient in the longer term because it gives the breeding pair relatively unchallenged access to resources. As is shown in the territory maps of golden eagles and great tits, territories space individuals apart and reduce intraspecific interactions. The size of the territory is related to the resources available within the defended area; larger territories are required when resources are poorer or widely dispersed. As is shown by the great tit example, when territory owners are removed, their areas are quickly occupied by birds previously displaced by competition.

Interspecific Competition (page 87)

1. The two species have similar niche requirements (similar habitats and foods). Red squirrels once occupied a much larger range than currently. This range has contracted steadily since the introduction of the greys. The circumstantial evidence points to the reds being displaced by the greys.

2. The greys have not completely displaced the reds. In areas of suitable coniferous habitat, the reds have maintained their numbers. In some places the two species coexist. **Note**: It has been suggested that the reds are primarily coniferous dwellers and extended their range into deciduous woodland habitat in the absence of competition.

3. Habitat management allows more effective long term population management *in-situ* (preferable because the genetic diversity of species is generally maintained better in the wild). Reds clearly can hold their own in competition with greys, provided they have sufficient resources. **Enhancing the habitat** preferred by the reds (through preservation and tree planting), assists their success as a competing species. Providing **extra suitable food plants** also enables the reds to increase their breeding success and maintain their weight through winter (thus entering the breeding season in better condition).

4. Other conservation strategies to aid red squirrel populations could include (any of): Captive breeding and release of reds into areas where they have been displaced, control/cull of grey squirrels (particularly in habitats suitable for reds), transfer of reds from regions where populations are successful to other regions of suitable habitat, supplementary feeding prior to the breeding season, public education to encourage red squirrels over greys.

5. (a) A represents the **realised niche** of *Chthamalus*.
   (b) When *Balanus* is removed from the lower shore, the range of *Chthalamus* extends into areas previously occupied by the *Balanus*; *Balanus* normally excludes *Chthamalus* from the lower shore.

6. Features will depend on the example chosen. Typical examples for two species that are proving to be invasive in many countries worldwide are described:
   (a) **Mallard duck** (*Anas platyrhynchos*):
   – Clutch size is relatively large compared with many other duck species.
   – Mallards are sexually aggressive and are capable of interbreeding with several native duck species where their ranges come to overlap (grey ducks in New Zealand, American black duck, the Florida mottled duck, and the endangered Hawaiian duck).
   – Mallards bully native duck species in competition for food and nest sites.
   – Mallards are adaptable in different environments.
   (b) **Mosquito fish** (*Gambusia affinis*):
   – Prolific breeders in many diverse environments.
   – Aggressive competitors for food.
   – Generalist feeders, and will prey on the eggs and larvae of native fish and frogs.
   – Reproductive strategies tuned to maximising reproductive output. Females are able to store sperm for extended periods and colonise environments without needing to meet a mate there.

Sampling Populations (page 89)

1. We **sample** populations in order to gain information about their abundance and composition. Sampling is necessary because, in most cases, populations are too large to examine in total.

2. (a) Random or systematic quadrat sampling.
   (b) Random or systematic point sampling (using a net or trap).
   (c) Line transect with point sampling (from low to high altitude). If time for sampling and analysis is not constrained, a belt transect using quadrats at regular intervals would provide the most information.

3. Information about the physical environment helps to explain species distributions. Certain species are usually associated with a particular suite of physical factors (e.g. preferred exposure, temperature, humidity etc) and if these are measured and known, more information on community patterns can be gathered.

4. As the vertical distance up the trunk increases, light intensity and temperature increase and humidity declines. With these changes in physical conditions there is consequent change in the vegetation from a diverse community of shade and moisture adapted moss species, to a community comprising just one species of (hardier) tree moss and various species of lichens, i.e. species more tolerant of the microclimate of lower humidity and higher light and temperature.

Quadrat Sampling (page 91)

1. Mean number of centipedes captured per quadrat:
   Total number centipedes ÷ total number quadrats
   30 individuals ÷ 37 quadrats
   = 0.811 centipedes per quadrat

2. Number per quadrat ÷ area of each quadrat
   $0.811 ÷ 0.08 = 10.1$ centipedes per m^2

3. Clumped or random distribution.

4. Presence of suitable microhabitats for cover (e.g. logs, stones, leaf litter) may be scattered.

Quadrat-Based Estimates (page 92)

1. Species abundance in plant communities can be determined by using quadrats and transects, and abundance scales are often appropriate. Methods for sampling animal communities are more diverse, and density is a more common measure of abundance.

2. **Size**: Quadrat must be large enough to be representative and small enough to minimize sampling effort.

3. **Habitat heterogeneity**: Diverse habitats require more samples to be representative because they are not homogeneous.

4. (a) and (b) any two of:
   - The values assigned to species on the scale are subjective and may not be consistent between users.
   - An abundance scale may miss rarer species and overestimate conspicuous ones.
   - The scale may be inappropriate for some habitats.
   - The semi-quantitative values assigned to the categories cover a range so results will lack precision.

Sampling a Leaf Litter Population (page 93)

It must be emphasised that the actual results for this practical are not particularly important. What is of value, is learning the value and limitations of this method, preferably **before** students are asked to carry it out in a field situation. The actual results will vary, depending upon the group's agreed upon criteria for inclusion of organisms in a given quadrat (i.e. when and how an organism is counted when it is partly inside a quadrat).
Note: Some leaves are almost completely obscured by invertebrates or have other leaves on top of them.

6. Typical results for samples used are:

	Direct count	A	B	C	D
Woodlice:	89	9.5	5	13	14.5
Centipedes:	3	0	0	1	1
Springtails:	6	0	3	0	0
False scorpions:	3	1	0	1	0
Leaves:	168	29	20.5	24.5	26.5

Class results will vary depending on counting criteria.

7. Typical results for calculated density are:

	Direct count	A	B	C	D
Woodlice:	2747	1759	926	2407	2685
Centipedes:	93	0	0	185	185
Springtails:	185	0	556	0	0
False scorpions:	93	185	0	185	0
Leaves:	5185	5370	3796	4537	4907

Note: Area of 6 quadrats = $(0.03 \times 0.03) \times 6 = 0.0054$ m^2
Area of total sample area = $0.18 \times 0.18 = 0.0324$ m^2

8. (a) **Problems** with sampling moving organisms: Once the quadrats have been laid, the animals moving from one quadrat to another risk being counted twice. **Solutions**: Physical barriers could be placed between each quadrat (what about the invertebrates directly underneath?). The entire area could be photographed for later analysis.
   (b) Data given above. Students should be made aware of the significance of extrapolating data from a small sample. The inclusion or exclusion of single individuals may have a large effect on the density figure, particularly where the species occur in low densities. Groups could combine data to see if they get a more representative sample.

Transect Sampling (page 95)

1. (a) With belt transects of any length (10 m or more), sampling (and sample analysis) using this method is very time consuming and labour intensive.
   (b) Line transects, although quicker than belt transects, may not be representative of the community. There may be many species which are present but which do not touch the line and are not recorded.
   (c) Belt transects use a wider strip along the study area and there is much less chance that a species will not be recorded.
   (d) It is not appropriate to use transects in situations involving highly mobile organisms.

2. To test whether or not the transect sampling interval was sufficient to accurately sample the community, the sampling interval could be decreased (e.g. from a sampling interval of every 1.5 m to an interval of every

0.25 m). If no more species are detected and the trends along the transect remain the same, then the sampling interval was adequate.

3.

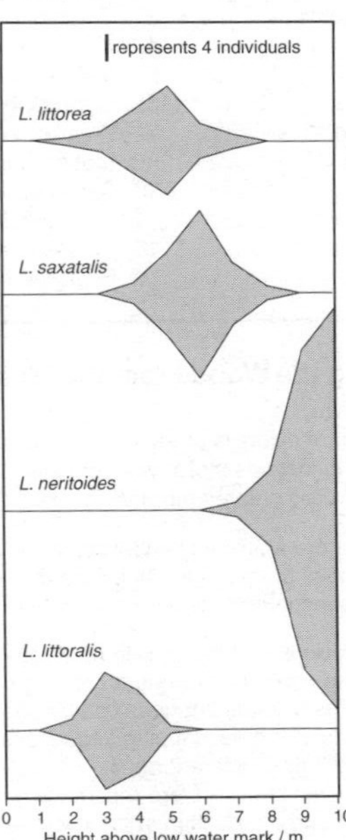

Mark and Recapture Sampling (page 97)

1. Results will vary from group to group for this practical. The actual results are not important, but it should serve as a useful vehicle for discussion of such things as sample size, variation in results between groups, and whether the method is a reliable way to estimate the size of a larger unknown group. Discussion could centre around what factors could be altered to make it a more reliable method (e.g. larger sample size, degree of mixing, increasing number of samples taken).

2. **Trout in Norwegian lake**:

Size of 1st sample:	109
Size of 2nd sample:	177
No. marked in 2nd sample:	57
Estimated total population:	109 x 177 ÷ 57 = 338.5

3. (a) Some marked animals may die.
   (b) Not enough time for thorough mixing of marked and unmarked animals.

4. (a) and (b) in any order (any two of):
   – Marking does not affect their survival.
   – Marked & unmarked animals are captured randomly.
   – Marks are not lost.
   – The animals are not territorial (must mix back into the population after release).

5. (a) Any animal that cannot move or is highly territorial (e.g. barnacle, tube worm, many mammals).
   (b) Unable to mix with unmarked portion of the population. Recapture at the same location would simply sample the same animals again.

6. (a)-(c) in any order:
   – Banding: leg bands of different colour on birds.
   – Tags: crayfish shell, fish skin, mammal ears.
   – Paint/dye used to paint markings in shell/fur.

7. Scientists hope to monitor fish growth and migratory movements. This will help them to establish the relationship between age and growth, which will help manage the population to prevent continued overfishing. Tracking of the tuna may also determine the location of further spawning grounds so that fish in these areas can be protected (presently there is only one known spawning ground). In addition to these data, researchers will find out more about the general biology of the tuna (e.g. data on feeding) which will help in future long term management of the species.

Sampling Animal Populations (page 99)

1. The tullgren funnel provides the best quantitative invertebrate sampling method. It is the only method that ensures that all of the invertebrates within a known area are removed, identified and counted to give a true reflection of population composition.

2. Pitfall traps rely on being placed in an area where the organisms are active. The traps take no account of clumped distributions or microhabitat preferences, and may overestimate densities in some areas and underestimate them in others.

3. (a) A large mesh size may fail to capture some smaller species or life stages. A fine mesh is apt to clog, and this reduces filtering efficiency so that much of the sampled volume is pushed out of the net instead of passing through it.
   (b) Mesh size should be fine enough to capture the species in which you are interested and it should be coarse enough to filter efficiently.

Primary Succession (page 100)

1. Glacial retreat, exposed slip, new volcanic island.

2. (a) Lichens and bryophytes (mosses and liverworts), as well as some hardy annual herb species, are often the first to colonise bare ground.
   (b) Chemically and physically erode the rock (producing the beginnings of a soil) and add nutrients by decay.

3. Climax communities tend to have greater biodiversity and a more complex trophic structure than early successional communities. A greater diversity of community interactions buffers the system against disturbances because there are many more organisms with different ecological roles able to compensate for losses from the system.

Secondary Succession (page 101)

1. **Primary succession** refers to the colonisation of regions where there is no preexisting community (e.g. rocky slope, exposed slip, new volcanic island). Changes in the community occur in stages until a climax community is reached. **Secondary succession** follows the interruption of an established climax community (e.g. logging, pasture reverting to bush).

2. Secondary succession proceeds more rapidly than is the case with primary succession because although the land is cleared, there is minimal or no loss of soil or seed stores. Many plants may still be able to grow despite the disturbance and the climax community will reestablish faster because nutrients are already available and seeds already laid down.

3. Events that result in secondary succession include forest fires, minor landslides, or storm or cyclonic damage, the fall of a canopy tree, flooding, or human induced land clearance.

4. (a) Selective logging causes gaps in the canopy with the loss of the large trees. This results in a chance for new seedlings to gain more light and grow up. These seedlings may be the same or different from the species that was/were removed. Forest composition is changed by becoming temporarily more open, with a loss of the dominant large trees.
   (b) Selective logging is considered (by some) to be preferable to clear felling because there is still a favourable environment in which seedlings can mature. Canopy gaps are created naturally by windfalls; selective logging is said to mimic this, so the gaps created *may* be beneficial for forest diversity and regeneration. Clear felling, in contrast, removes all forest from the area leaving an environment unfavourable for the regeneration of existing forest species but favourable for weeds. forest structure and diversity.

5. (a) A **deflected succession** refers to a succession that is deflected from its natural course by human intervention. The plagioclimax that develops is different from the one that would have developed if the intervention had not occurred.
   (b) Many human-modified landscapes (e.g. agricultural lands) are managed (by burning, grazing, mowing) with the express purpose of preventing the establishment of a natural climax community. These communities are quite distinct from those that would naturally develop if the land were left alone.

Wetland Succession (page 103)

1. (a) Low evaporation rates   (b) High rainfall

2. Both these conditions are found at high latitudes and/or high altitudes.

3. Land drainage threatens bog ecosystems because it dries out the soil and allows acid intolerant species to invade and compete with the natural bog species.

4. *Sphagnum* lowers the pH of the surrounding soil and hampers the establishment of the acid intolerant species typical of swamps and fens. Acid tolerant (bog) species can then become established.

Key Terms: Crossword (page 104)

Answers Across	Answers Down
1. Habitat	2. ACFOR
7. Photosynthesis	3. Thylakoid
9. Niche	4. ATP
10. Carbon	5. Interspecific
11. RuBisCo	6. Light dependent
12. Producer	7. Population
15. Quadrat	8. Biotic factor
17. Lincoln	13. Calvin cycle
18. Succession	14. Chlorophyll
20. Abiotic factor	16. Abundance
21. Consumer	19. Ecosystem

Separating the Wood From the Trees (page 106)

1. The wood of dicots is more variable than conifers because it contains more types of cells. The greater variety of cells creates more variability in the wood.

2. Tree rings are analysed by looking at the number and the thickness of the rings. Rings are thicker during the growing season and thinner during winter.

3. (a) The thickness of the growth rings can indicate growing conditions during a growing season for that tree. Past climate changes can be inferred from the analysis of the overlapping sequences of growth rings of many different trees.
   (b) Short lived stressors are recorded by reduced growth and a reduction in the thickness of the seasonal growth ring. Scientists can analyse variations in annual rings to determine the extent and severity of short lived environmental events. These will be distinct and can be correlated against evidence from other sources.

Global Warming (page 107)

1. (a) Carbon dioxide: 30.9% increase
   (b) Methane: 111.0% increase
   (c) Nitrous oxide: 11.6% increase

2. Higher temperatures will result in (main points):
   – Accelerated melting of the Greenland ice sheets.
   – Releases of terrestrial carbon from thawing permafrost.
   – Release of CH_3 from hydrates in coastal sediments.
   – Expansion in ocean volume. Extra water (previously locked up in glaciers) will add to this increase.
   – Arctic melt season begins earlier and ends later.
   – Many Arctic species (e.g. polar bears) are at risk, especially those that inhabit the low lying areas near the sea and those that rely on cold weather conditions. Glaciers and ice sheets provide floating oases for wildlife (e.g. for rest when moving between bergs) and disappearing habitat is already forcing species such as polar bears and seals to make longer, riskier swims.
   – Reductions in sea ice reduce habitat for marine zooplankton (e.g. amphipods) and shifts species assemblages of phytoplankton. Both of these consequences alter the dynamics and resilience of marine food chains.
   – Tundra ecosystems will also be adversely affected as permafrost loses depth and species from all trophic levels are lost or replaced.

- Positive feedback loops are inherent in all these changes as multiple processes accelerate ice loss.

3. Increased levels of carbon dioxide, methane, and nitrous oxides act as additional blankets around the Earth, allowing the sun's energy to reach the Earth's surface, but preventing the heat escaping. This means that the Earth slowly heats up.

4. Any one of:
   - Most national governments have signed and ratified the Kyoto Protocol aimed at combating greenhouse gas emissions.
   - Measures are in place to reduce or reverse future warming or to adapt to its expected consequences (mitigation or adaptation). Examples include reducing dependence on fossil fuels and long term planning to prepare for sea level rise and more severe weather events etc.)
   - Carbon credit schemes provide a tradable permit scheme to create a market for reducing greenhouse emissions by giving a monetary value to the cost of polluting the air.

Effects of Global Warming (page 109)

1. Global warming will result in an increased frequency of weather extremes (floods, droughts etc) and a loss of land as coastal areas are inundated. Erosion rates may also increase as a result. Glacial retreats will reduce water supplies and snow lines will increase in altitude. Climate changes may shift the governing physical environment in certain regions (and consequently cause a shift in predominant vegetation). Ocean pH will also fall as a result of CO_2 absorption (again, with consequent changes in biotic communities).

2. (a) In general, crop growing ranges may shrink, expand, or shift. Crop plants may be affected more by higher night temperatures than by higher daytime temperatures. High night temperatures affect the ability of some crop plants such as rice to set seed and fruit. This will cause a reduction in the harvest, and a decrease in the amount of seed available for subsequent plantings.
   (b) Farmers may adjust by planting different crops in some areas, e.g. crops that are able to grow and set seed in the higher temperatures. New strains of crop plants may be able to be developed for the higher temperatures.

3. Fossil evidence from studies of browse damage show a significant increase in the number of damaged leaves coincident with temperature peaks just under 56 mya. If insect populations respond similarly to future temperature increases, crop damage could increase.

4. (a) Migratory birds in the northern hemisphere are now not travelling as far south during the winter months (as higher latitudes become more hospitable) and they are making their migrations north up to two weeks earlier than usual.
   (b) Migratory birds may arrive at feeding grounds before the main food supply is ready. Plants with daylength-dependent flowering may not yet be flowering and, as a consequence, insects (and seeds and fruits) may not be in the abundance required to feed the migrants. In addition, the distribution of food resources may remain the same, but the birds are not migrating as far south and may be disconnected from their winter food supplies.

5. As air temperatures rise, so too does the snow line in alpine areas. Animals living on or above the snow line will be forced into smaller areas. If they are unable to move to higher latitudes where the snow line is lower, it is inevitable that they will become extinct in their native ranges as they run out of food and space.

The Effect of Temperature (page 111)

1. Species distribution of *Rana* is closely related to water temperature (mating and embryonic development for each species occurs within a certain temperature range). Increasing global temperatures may result in a northwards shift of some species as their preferred water temperature shifts. Those frogs that are furthest north (*R. sylvatica*) may end up with a reduced range, while those further south (*R. clamintans*) may increase their range, depending on available habitat.

2. (a) Increased temperature speeds up the development rate in both species but, in *C. dubia*, development rates are always faster at any given temperature.
   (b) The ability of *C. dubia* to develop faster at lower temperatures gives it a competitive advantage over *C. pulchella*. *C. dubia* populations can expand early in the season without competition from *C. pulchella*, whose developmental rates lag behind at lower temperatures.

3. (a) Time to first reproduction decreases with increasing temperature. At 13°C there is a 1.5 day advantage to *C. dubia*, which reaches maturity much earlier, but at 23°C this advantage is much less.
   (b) Egg development time decreases for both species as temperature increases. However, *C. dubia* develops faster than *C. pulchella* at all temperatures presented in this study. *C. dubia* has a significant advantage at temperatures <15°C.
   (c) A temperature increase would reduce the competitive advantage of *C. dubia* over *C. pulchella* (namely earlier reproduction, development, and population establishment). The two species are likely to be competing for the same resources (small edible algae) at the same time, so both populations may be adversely affected.

4. Polar bears are forced to swim longer distances to obtain prey, and to return to the mainland for birthing. The ice sheet becomes thinner earlier in the season, and will not support their weight, so they must return to the mainland earlier. As a result, they have fewer energy reserves to get them through winter, reproductive rates are lower, as are juvenile survival rates. Their population is in decline.

The Global Warming Debate (page 113)

1. Climate change predictions involve many complicated parameters and will never be fully accurate because they are based on scenarios in which parameters can only ever be estimated. Some parameters are simplified or ignored in order to make the models workable. It is also difficult to predict future human actions, and the effect of these actions on climate change. Therefore, scientists can only suggest a likely course of events; they can not provide a definite outcome.

2. The climate change models are updated as new information is added to the model. In addition, advancements in computer modelling algorithms help refine the predictions. These changes enable scientists to produce more accurate models of climate change.

3. Parties whose activities contribute to global warming (e.g. producers and consumers of coal and oil) are likely to try to downplay the data, or provide alternative data because making changes to existing business or production processes will cost them money. The opposite is true of companies that stand to make a financial gain from the global warming debate, e.g. a company manufacturing biofuels or bioplastics may emphasise data linking fossil fuel consumption with global warming to promote themselves as a greener alternative, but they may not mention research linking biofuels to the global food shortage.

4. The media do not always provide a balanced argument about climate change. Some media outlets have very strong political affiliations and present only one side of the debate. The public view on climate change could be skewed by media presentation unless the accounts provided are impartial. In addition, the public have become weary of climate change; many people switch off to the debate if the media coverage is unrelenting.

5. While electric cars will reduce CO_2 emissions, they will require electricity to recharge their battery. If the electricity is produced using a non-renewable source (e.g. coal), the reduction in transport CO_2 emissions will be partly negated by the power production required to charge the battery. Some biofuels are produced from food crops. This reduces available food for consumption.

Temperature and Enzyme Activity (page 115)

1. All enzymes have an optimum temperature range at which they work best. An increase in environmental temperature may put a poikilothermic organism (and its enzymes) outside its optimal range. Enzyme activity may be reduced to suboptimal levels as a result.

2. Enzymes are usually part of complicated metabolic pathways and their activity relies on substrate availability and usually the action of other enzymes or cofactors as well. In addition, there are unknown variables; enzymes studied in isolation often react differently to enzymes within an organism or whole system, where there may be buffers against change.

Extinction or Evolution (page 116)

1. (a) Adaptation by evolution is the result of changes to a population's gene pool. These changes are the result (in part) of natural selection. Phenotypic plasticity causes alteration in phenotype in response to environmental change, but there are no changes to the gene pool.
   (b) Adjustment to the environment involves changes within individuals that are within the phenotypic range of the organism and may happen relatively quickly. Adaptive changes involve changes to the gene pool and take place over many generations.

2. Some species have a narrower physiological tolerance range and/or more limited phenotypic plasticity than others. These species will have limited capacity to tolerate change or make the phenotypic adjustments necessary to survive. For immobile organisms (e.g. plants), the risk may be greater because they cannot move to another location. It their offspring/seeds cannot be spread to a more favourable environment, and the individuals cannot survive and propagate where they are, then extinction is more likely.

Mutation and Natural Selection (page 117)

1. People who are heterozygous for the sickle cell gene are somewhat affected by sickle cell anemia but have considerable resistance to malaria which is widespread in the region. This heterozygous advantage maintains the mutant allele at a relatively stable frequency in the population despite its deleterious effects. The stable coexistence of both the sickle cell allele and the normal allele exemplifies a balanced polymorphism.

Evolution of Drug Resistance (page 118)

1. Antibiotic resistance refers to the resistance bacteria show to antibiotics that would normally inhibit their growth. In other words, they no longer show a reduction in growth response in the presence of the antibiotic.

2. (a) Antibiotic resistance arises in a bacterial population as result of mutation. Some bacteria can also acquire these changes in DNA (conferring resistance) by transfer of genes between bacteria by conjugation (horizontal evolution).
   (b) Resistance can become widespread as a result of (1) transfer of genetic material between bacteria (horizontal evolution) or by (2) increasing resistance with each generation a result of natural selection processes (vertical evolution). In the latter case, the antibiotic provides the environment in which selection for resistance can take place. It is exacerbated by overuse and misuse of antibiotics.

3. Widespread antibiotic resistance has implications for the treatment and control of what have been, in the past, quite easily treated diseases. Tuberculosis is one good example. Historically, it was effectively treated with antibiotics, but complacency over its control has lead to increasing multiple drug resistance in the *Mtb* population and a resurgence in the number of TB cases. This has huge implications for public health because more people live with (resistant forms of) the disease and spread it to more people as a result. In addition, the costs associated with treating TB are now also much higher. In general, increasing resistance increases the costs lowers the efficacy of treating disease.

Antigenic Variability in Pathogens (page 119)

1. (a) The viral genome is contained on 8 short, loosely connected RNA segments. This enables ready exchange of genes between different viral strains and leads to alteration on the protein composition of the H and N glycoprotein spikes.
   (b) The body's immune system acquires antibodies to the H and N spikes (antigens) on the viral surface, but when different variants arise they are not recognised nor detected by the immune system (there is no immunological memory for the newly appearing antigens).

2. An antigenic shift represents the combination of two or more different viral strains in a new subtype with new properties and no immunological history in the population. In contrast, antigenic drifts are much smaller changes that occur continually over time and to which small adjustments (to the flu vaccine or to immune response) are sufficient to provide resistance.

Insecticide Resistance (page 120)

1. (a) and (b) any two in any order:
- Insect populations tend to reproduce very quickly (generation times are short) and so the chances of a mutation conferring resistance (1) arising and (2) being passed on are greater.
- Partial resistance can arise at several levels (e.g. behavioural, mechanical, biochemical) and, through sexual reproduction in a given selective environment, offspring can require a sufficient number of mechanisms to develop full resistance.
- Applications of insecticide do not always reach their intended target in the correct dosage and so do not kill 100% of the population, allowing surviving individuals to reproduce and pass on even slight resistances.

2. The insecticide application may not kill off the entire insect population. Those that survive will pass on any resistance they have to the next generation. Periodic insecticide applications act as a selection agent by allowing only the most resistant insects to survive and pass on their genes.

3. Resistance to synthetic insecticides has a number of implications. In order to maintain the kill rate of the insecticide, farmers often increase the amount/potency/ toxicity of insecticide applied. This leaves more residue on the plant and so is potentially more dangerous to humans. Crops may have to be withheld longer before going to market and so cause loss of income to the farmer. Increasing resistance among insect vectors of disease is also a problem for human populations (e.g. resistance in mosquito vector for malaria), reducing the options for controlling the spread of disease through vulnerable human populations.

Selection for Skin Colour in Humans

(page 121)
1. (a) Folate is essential for healthy neural development. Note: A deficiency causes (usually fatal) neural tube defects (e.g. spina bifida).
   (b) Vitamin D is required for the absorption of dietary calcium and normal skeletal development. Note: A deficiency causes rickets in children or osteomalacia in adults. Osteomalacia in pregnancy can lead to pelvic fractures and inability to carry a pregnancy to term.

2. (a) Skin cancer normally develops after reproductive age. Protection against it provides no reproductive advantage and so no mechanism for selection.
   (b) The new hypothesis for the evolution of skin colour links the skin colour-UV correlation to evolutionary fitness (reproductive success). Skin needs to be dark enough to protect folate stores from destruction by UV and so guard against fatal neural defects in the offspring. However it also needs to be light enough to allow enough UV to pentrate the

skin on order to manufacture vitamin D for calcium absorption. Without this, the female skeleton cannot successfully support a pregnancy. Because these pressures act on individuals both before and during reproductive age they provide a mechanism for selection. The balance of opposing selective pressures determines eventual skin coloration.

3. Women have a higher requirement for calcium during pregnancy and lactation. Calcium absorption is dependent on vitamin D, making selection pressure on females for lighter skins greater than for males.

4. The Inuit people have such abundant vitamin D in their diet that the selection pressure for lighter skin (for UV absorption and vitamin D synthesis) is reduced and their skin can be darker.

5. (a) Higher risk of rickets or (the adult equivalent) osteomalacia due to low UV absorption.
   (b) The simplest option to avoid these problems is for these people to take dietary supplements to increase the amount of vitamin D they obtain.

Analysis of a Squirrel Gene Pool (page 123)

1. Graph of population changes:

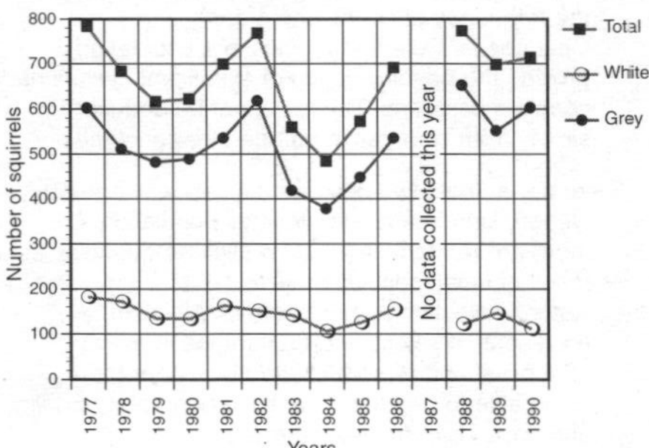

(a) 784 to 484 = 61% fluctuation
(b) Total population numbers exhibit an oscillation with a period of 5-6 years (2 cycles). Fluctuations occur in both normal grey and albino populations.

2. Graph of genotype changes:

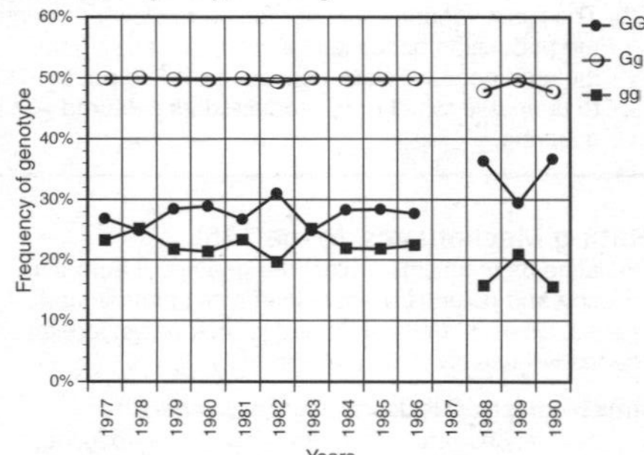

(a) GG genotype: Relatively constant frequency until the last 3-4 years which exhibit an increase. Could also argue that there is an increase over the total sampling period.

(b) Gg genotype: Uniform frequency.
(c) gg genotype: Relatively constant frequency until the last 3-4 years which exhibit a decline. Could also argue that there is a decrease over the total sampling period.

3. Graph of allele changes:

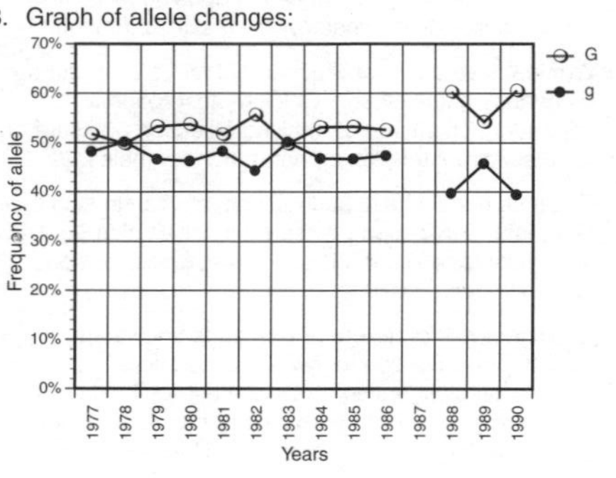

(a) Frequency of G: Increases in the last 3-4 years.
(b) Frequency of g: Decreases in the last 3-4 years.

4. (a) The frequency of alleles graph (to a lesser extent the frequency of genotypes graph)
    (b) Changes in allele frequencies in a population provide the best indication of significant evolutionary changes occurring. These cannot be deduced simply from changes in numbers or genotypes.

5. There are at least two possible causes (any one of):
    – Genetic drift in a relatively small population, i.e. there are random changes in allele frequencies as a result of small population size.
    – Natural selection against albinos. Albinism represents a selective disadvantage in terms of survival and reproduction (albinos are more vulnerable to predators because of greater visibility and have lower fitness generally).

Black is the New Grey (page 125)

1. (a) The mutation for black coat colour shows incomplete dominance, so any mating involving a black squirrel will result in at least 50% dark coloured offspring (either brown black or jet black).
    (b) The black variant would spread more slowly through the population because two squirrels carrying the mutated gene for black coat colour would need to be mated together to produce dark coloured offspring.

Isolating Mechanisms (page 126)

1. Isolating mechanisms protect the gene pool from the diluting and potentially adverse effects of introduced genes. Species are finely tuned to their niche; foreign genes will usually reduce fitness.

2. (a) Geographical isolation physically separates populations (and gene pools) but, if reintroduced, the two populations could potentially interbreed, i.e. reproductive isolation may not have occurred.
    (b) Geographical isolation enables populations to diverge in response to different selection pressures and (potentially) develop reproductive isolating

mechanisms. Reproductive isolation won't generally occur in a populations in which there is gene flow (unless by special events such as polyploidy).

3. Geographical isolation physically separates populations (gene pools) so there is no gene flow between them. Ecological isolation arises as a result of different preferences in habitat or behaviour even though the populations occupy the same geographical area.

Reproductive Isolation (page 127)

1. (a) Postzygotic: hybrid breakdown
    (b) Prezygotic: structural
    (c) Prezygotic: temporal
    (d) Postzygotic: hybrid inviability

2. They are a secondary backup if the first isolating mechanism fails. The majority of species do not interbreed because of prezygotic mechanisms. Postzygotic mechanisms are generally rarer events.

Allopatric Speciation (page 129)

1. Interspecific competition, intraspecific competition (this is the strongest). New habitat becomes available (loss of geographical barrier).

2. (a) Plants move by dispersal of their seeds (water borne, carried vast distance by winds, animals carry them in their fur/feathers or in their gut to be deposited with their faeces).
    (b) Gene flow between the parent population and dispersing populations is regular. Gene flow refers to the movement of genes, via gametes, from one population to another.

3. Orogeny or plate tectonics (continental drift).

4. Ice ages, glacials, or interglacials.

5. Mountains form. Rivers form or change course. Deserts. Isthmus or land bridge forms (land separating two seas, e.g. Central America). Ice sheets advance. Sea or ocean develops (due to continental drift).

6. Physical (or geographical) barriers prevent the movement of organisms between populations and thereby restrict or prevent gene flow between separated populations. Once this happens, populations may diverge genetically on either side of the physical barrier.

7. (a) Selection pressures on gene pools (any four of): Various climatic factors (e.g. temperature, pH, salinity, wind exposure), competition, predation, disease and parasitism, quality of the food resource.
    (b) Some individuals will have allele combinations (and therefore a phenotype) that better suits the unique set of selection pressures at a given location. Over a period of time (generations) certain alleles for a gene will become more common in the gene pool, at the expense of other less suited alleles.

8. Reproductive isolating mechanisms:
    (see pages 122-124 of the workbook for full descriptions).
    (a) **Prezygotic** (prevent fertilisation from taking place): Spatial, ecological, temporal, behavioural, structural, gamete mortality.
    (b) **Postzygotic** (fertilisation occurs, but the result is unsuccessful, to varying degrees): Zygote mortality,

reduced fertility, hybrid breakdown.

9. **Allopatry** refers to the situation where populations within the species have separate areas of geographical range. **Sympatry** refers to the situation where populations within the species inhabit the same or overlapping geographical areas.

Small Flies and Giant Buttercups (page 131)

1. When the original species of drosophilidae arrived on the Hawaiian islands it found many new unoccupied niches into which it expanded, leading to an **adaptive radiation**.

2. The fruit flies are of interest because there are so many closely related species within a small area an speciation has been (and is) relatively frequent. The flies also have a relatively simple genome, making genetic studies relatively easy.

3. In general the oldest species of flies are found on the oldest islands. As islands appeared out of the sea the flies spread to new environments and diversified, giving rise to newer species.

4. Buttercups living in alpine areas periodically have their habitats reduced and their range restricted during periods of climatic warming. This restricts gene flow and leads to speciation. Periods of cooling allow for the expansion of their range and movement to new environments as well as hybridisation to form new species. Repeated many times, these cycles lead to a large range of species.

DNA Homologies (page 132)

1. The similarity of DNA from different species can be established in a rudimentary way by measuring how closely single strands from each species mesh together. The more similar the DNA, the harder it is to separate them. These studies have confirmed most evolutionary relationships guessed at from anatomical comparisons.

2. (a) Chimpanzee   (b) Galago

3. (a) 7 – 8   (b) Approx. 12

4. 45 million years ago.

Protein Homologies (page 133)

1. Chimpanzees and gorillas have virtually identical amino acid sequences to humans for some proteins.

2. (a) They play a crucial role in the respiratory pathway. Most changes are likely to be deleterious so they change very little over time.
   (b) Such proteins are good candidates for use in establishing homologies because the few changes that are retained through time are likely to be meaningful, i.e. represent major divergences in evolutionary lines.

3. Any of: The rate of change must be calibrated against material evidence (e.g. fossils) for firm conclusions to be made. The functions of the protein may change over time. Clock may run at a different rate in different species.

Hox Genes (page 135)

1. **Evo-devo**, or evolutionary developmental biology, is the study of the origin and evolution of embryonic development in animals. It traces the role of genes in developmental processes and looks at how modifications of these processes may lead to the evolution of novel features.

2. Homeotic gene sequences, such as the *Hox* genes, make up a genetic tool kit for animal development. These genes act as genetic switches (transcription factors), turning other genes on or off in the course of development. For example, the *Hox* genes determine the positioning of body parts in organisms as diverse as *Drosophila* and mice. The same genes are present in essentially all animals, including humans, and mutations in these genes can have a profound effect on morphology.

3. Evo-devo is important as evidence for evolution because it explains how novel forms may arise without attributing these changes to the evolution of new genes. Instead natural selection associated with gene regulation can be explained as a mechanism for the evolution of novel forms.

4. Changes in gene expression can bring about changes in morphology, e.g. in differences in neck length in vertebrates and eyespot development on butterfly wings. Neck length is controlled by expression of the *Hox c6* gene. Where it is expressed marks the boundary between neck and trunk vertebrae, e.g. in snakes the boundary is shifted forward to the base of the skull and results in no neck, while in geese the boundary is shifted backwards and a long neck results. Similarly, the development of eyespots on butterfly wings is controlled by switches in the *Distal-less* gene, whose expression results in a range of spots from virtually all eyespot elements expressed to very few.

Peer Review Process (page 137)

1. The peer review process uses experts in a particular scientific field to validate new research. It ensures that good methodology has been used, and that the results and conclusions are valid and appropriate.

2. Reviewers assume the work submitted is valid, so are not looking for false data. In addition, reviewers do not have access to the raw data, only the transformed data, so it can be very difficult to tell that data has been manipulated falsely or fabricated.

3. Double blind reviews allow a thorough critique without fear of academic retribution by a disgruntled submitter. It also stops reviewers from being swayed by the reputation of a scientist or institution. A single blind review has only the first element.

Key Terms: Mix and Match (page 138)

Allele (Q), Allele frequency (W), allopathic (N), Carbon dioxide (M), Climate change (V), Dendrochronology (J), Enzyme (U), Evolution (T), Extinction (P), Gene pool (C), Global warming (R), Greenhouse effect (D), Greenhouse gas(B), Homology (H), Hox gene (F), Isolating mechanisms (A), Methane (O), Molecular clock (I), Mutation (X), Natural selection (L), Peer review (G), Postzygotic (S), Prezygotic (E), Reproductive isolation (K).

The Genetic Code (page 140)

1. This exercise demonstrates the need for a 3-nucleotide sequence for each codon and the resulting degeneracy in the genetic code.

Amino acid	Codons						No.
Alanine	GCU	GCC	GCA GCG				4
Arginine	CGU	CGC	CGA CGG	AGA AGG			6
Asparagine	AAU	AAC					2
Aspartic Acid	GAU	GAC					2
Cysteine	UGU	UGC					2
Glutamine	CAA	CAG					2
Glutamic Acid	GAA	GAG					2
Glycine	GGU	GGC	GGA GGG				4
Histidine	CAU	CAC					2
Isoleucine	AUU	AUC	AUA				3
Leucine	UAA	UUG	CUU CUC	CUA CUG			6
Lysine	AAA	AAG					2
Methionine	AUG						1
Phenylalanine	UUU	UUC					2
Proline	CCU	CCC	CCA CCG				4
Serine	UCU	UCC	UCA UCG	AGU AGC			6
Threonine	ACU	ACC	ACA ACG				4
Tryptophan	UGG						1
Tyrosine	UAU	UAC					2
Valine	GUU	GUC	GUA GUG				4

2. (a) 16 amino acids
   (b) Two-base codons (eg. AT, GG, CG, TC, CA) do not give enough combinations with the 4-base alphabet (A, T, G and C) to code for the 20 amino acids.

3. Many of the codons for a single amino acid vary in the last base only. This would reduce the effect of point mutations, creating new and potentially harmful amino acid sequences in only some instances. **Note:** Only 61 codons are displayed above. The remaining three are **terminator** codons (labelled 'STOP' on the table in the workbook). These are considered the 'punctuation' or controlling codons that mark the end of a gene sequence. The amino acid **methionine** (AUG) is regarded as the 'start' (initiator) codon.

Gene Expression (page 141)

1. In **prokaryotic** gene expression, RNA is translated into protein almost as fast as it is transcribed; there is no nucleus therefore no separation of the transcription and translation processes. The presence of introns in a prokaryotic genome would interfere with protein function because there would be no time between transcription and translation to splice them out.
   In contrast, **eukaryotic** gene expression involves production of a primary RNA transcript from which the introns are removed. There is sufficient time for this to take place because transcription and translation occur inside and outside the nucleus respectively. **Note**: Evidence in support of this: Prokaryote DNA consists almost entirely of protein-coding genes and their regulatory sequences with very little non-protein coding DNA. Eukaryotic DNA comprises large amounts of non-protein coding sequences, much of which we now know codes for functional (regulatory) RNAs.

2. In the new view of eukaryotic gene expression, the so-called "junk DNA", traditionally regarded as non-functional, is considered to play a necessary role in the eukaryote cell. In fact, there is a good correlation between the amount of non-protein coding DNA a species has and its complexity. Although this DNA does not code for proteins, its does code for RNA molecules with regulatory functions (some of these involving regulation of the genome itself). What's more, in the new view of gene expression, not all of the exonic RNA is translated into protein; some contributes to regulatory microRNAs. In the old view, all exonic RNA was thought to code for proteins.

3. The one gene-one protein model is still appropriate for prokaryotes because, in the absence of introns, RNA transcripts are translated directly into proteins. Because of the small size of prokaryotic genomes and their one site of protein synthesis, there is very little non-protein coding DNA present.

Transcription (page 143)

1. mRNA carries a copy of the genetic instructions from the DNA in the nucleus to ribosomes in the cytoplasm. The rate of protein synthesis can be increased by making many copies of identical mRNA from the same piece of DNA.

2. (a) AUG   (b) UAA, UAG, UGA

3. (a) AUG AUC GGC GCU AAA
   (b) AUG UUC GGA UAU UUU

Translation (page 144)

1. AUG AUC GGC GCU AAA

2. (a) 61
   (b) There are 64 possible codons for mRNA, but three are terminator codons. 61 codons for mRNA require 61 tRNAs each with a complementary codon.

Post Transcriptional Modification (page 145)

1. Primary mRNA can be spliced and combined in many different ways. This allows more proteins to be produced than the number of genes in the human genome. Post translational modification also allows for more protein variations to be produced.

2. There are four main ways in which mRNA can be modified to code for different proteins. (1) Exon skipping can leave out one or more parts of the mRNA. (2) Intron retention retains parts of the primary mRNA that would normally be left out. (3) Mutually exclusive exons can produce two alternative proteins, and (4) some exons have more than one binding site, which produces exons of different lengths and different proteins.

3. Being able to modify the mRNA allows an organism to produce alternative proteins without needing an entirely new gene to code for it. This ensures that the organism's genome is not too large. For example the *Dscam* gene in *Drosophila melanogaster* can produce more than 38 000 proteins from the one gene.

Post Translational Modification (page 146)

1. (a) Modifications include: (1) Cleaving the chain to form smaller molecules, which then combine to form the functional protein. (2) Adding carbohydrate or phosphate groups to alter the structure or destination of the protein. (3) Attaching lipids to anchor the protein in the cell membrane. (4) Degrading the protein chain once it has served its purpose.
    (b) Changes to the polypeptide determine the protein's structure and function, or determine where the protein is to transported. Without these modifications the protein may not function correctly.

2. Protein orientation in the membrane is usually critical to the functional role of the protein (e.g. intercellular recognition or transport).

Protein Synthesis Summary (page 147)

1. A DNA
    B Free nucleotides
    C RNA polymerase enzyme
    D mRNA
    E Nuclear membrane
    F Nuclear pore
    G tRNA
    H Amino acids
    I Ribosome
    J Polypeptide chain

2. **8** tRNA molecule is recharged with another amino acid of the same type, ready to take part in protein synthesis.
    5 tRNA molecule brings in the correct amino acid to the ribosome.
    1 Unwinding the DNA molecule.
    3 DNA rewinds into double helix structure.
    6 Anti-codon on the tRNA matches with the correct codon on the mRNA and drops off the amino acid.
    7 tRNA leaves the ribosome.
    4 mRNA moves through nuclear pore in the nuclear membrane to the cytoplasm.
    2 mRNA synthesis: Nucleotides added to the growing strand of messenger RNA molecule.

3. The process by which the cell (and therefore the organism) manufactures proteins that may be used as building materials or controlling chemicals.

4. Ribosomal RNA (rRNA), as part of ribosomal structure. Transfer RNA (tRNA). Messenger RNA (mRNA).

5. (a)-(c) in any order:
    – RNA is single stranded, DNA is double stranded.
    – RNA contains the pentose sugar ribose, whereas DNA contains deoxyribose.
    – DNA takes up the twisted double helix whereas RNA does not.
    – RNA has had non-coding codons (introns), which were present in the DNA, removed.
    – Uracil replaces thymine in the code for mRNA.
    – It is mRNA that carries the message of the DNA coding strand in protein synthesis.

6. (a) Transcription   (b) Translation

7. (a) UUU to UUC: This change to the third base still codes for the amino acid phenylalanine.
    (b) UUU to UUA: This change to the third base changes the code for the amino acid from phenylalanine to leucine.
    (c) UUU to UUA will change the protein produced.

Gel Electrophoresis (page 149)

1. Gel electrophoresis separates mixtures of molecules (proteins, nucleic acids) on the basis of size and other physical properties.

2. (a) The frictional (retarding) force of each fragment's size (larger fragments travel more slowly than smaller ones).
    (b) The strength of the electric field (movement is more rapid in a stronger field).
    Teacher's note: The temperature and ionic strength of the buffer can be varied to optimise separation.

3. The gel is full of pores (holes) through which the fragments must pass. Smaller fragments pass through these pores more easily (with less resistance and therefore faster) than larger ones.

Interpreting Electrophoresis Gels (page 150)

1. Use the mRNA table on the activity page: *The Genetic Code*, to determine the amino acid sequence.
    (a) Synthesised DNA:
    CGT AAG TAC TTG ATC AGA
    GCT CTT CGA AAA TCG
    (b) DNA sample:
    GCA TTC ATG AAC TAG TCT
    CGA GAA GCT TTT AGC
    (c) mRNA:
    CGU AAG UAC UUG AUC AGA
    GCU CUU CGA AAA UCG
    (d) Amino acids:
    Arg Lys Tyr Leu Iso Arg Ala Leu Arg Lys Ser

2. (a) ATG ATC GGC GCT AAA TGT TAA
    (b) ATG CGG AAT TTC CCG GCT TAG
    (c) DNA replication.

3. 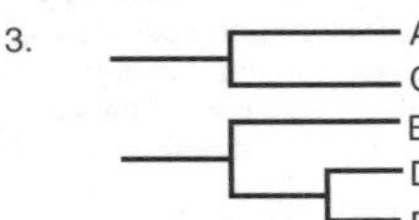

Polymerase Chain Reaction (page 151)

1. To produce large quantities of 'cloned' DNA from very small samples. Large quantities are needed for effective analysis. Small quantities are often unusable.

2. A double stranded DNA is heated to 98°C for 5 min, causing the two strands to separate. Primers, free nucleotides, and DNA polymerase are added to the sample. The sample is then cooled to 60°C for a few minutes, and the primers anneal to the DNA strands. The sample is incubated and complementary strands are created (by the DNA polymerase) using each strand of the DNA sample as a template. (Incubation temperature depends on the particular polymerase used; it is ~72°C for heat tolerant Taq polymerase, but lower for other polymerases). The process is repeated about 25 times, each time the number of templates doubles over the previous cycle.

3. (a) Forensic samples taken at the scene of a crime (e.g. hair, blood, semen).
    (b) Archaeological samples from human remains.
    (c) Samples taken from the remains of prehistoric

organisms preserved in ice, mummified, preserved in amber, tar pits etc.

4. This exercise can be done on a calculator by pressing the 1 button (for the original sample) and then multiplying by 2 repeatedly (to simulate each cycle).
   (a) 1024
   (b) 33 554 432   (33.5 million)

5. (a) It would be amplified along with the intended DNA sample, thereby contaminating the sample and rendering it unusable.
   (b) Sources of contamination (any two of):
   Dirty equipment (equipment that has DNA molecules left on it from previous treatments).
   DNA from the technician (dandruff from the technician is a major source of contamination!)
   Spores, viruses and bacteria in the air.
   (c) Precautions to avoid contamination (any two of):
   Using disposable equipment (pipette tips, gloves).
   Wearing a **head cover** (disposable cap).
   Use of **sterile procedures**.
   Use of **plastic disposable tubes with caps** that seal the contents from air contamination.

6. (a) and (b) any of the following procedures require a certain minimum quantity of DNA in order to be useful: DNA sequencing, gene cloning, DNA profiling, transformation, making artificial genes. Descriptions of these procedures are provided in the workbook.

DNA Profiling Using PCR (page 153)

1. STRs (microsatellites) are non-coding nucleotide sequences (2-6 base pairs long) that repeat themselves many times over (repeats of up to 100X). The human genome has numerous different STRs; equivalent sequences in different people vary considerably in the numbers of the repeating unit. This property can be used to identify the natural variation found in every person's DNA since every person will have a different combination of STRs of different repeat length.

2. (a) **Gel electrophoresis**: Used to separate the DNA fragments (STRs) according to size to create the fingerprint or profile.
   (b) **PCR**: Used to make many copies of the STRs. Only the STR sites are amplified by PCR, because the primers used to initiate the PCR are very specific.

3. (a) Extract the DNA from sample. Treat the tissue with chemicals and enzymes to extract the DNA, which is then separated and purified.
   (b) Amplify the microsatellite using PCR. Primers are used to make large quantities of the STR.
   (c) Run the fragments through a gel to separate them. The resulting pattern represents the STR sizes for that individual (different from that of other people).

4. To ensure that the number of STR sites, when compared, will produce a profile that is effectively unique (different from just about every other individual). It provides a high degree of statistical confidence when a match occurs.

Forensic Applications of DNA Profiling
(page 155)

1. Lane A acts a control or calibration lane containing

fragments of DNA of known length.

2. Profiles of everyone involved must be completed to compare their DNA to any DNA found at the scene and therefore eliminate (or implicate) them as suspects.

3. The alleged offender is not guilty. The alleged offender's DNA profile does not appear in the DNA collected at the crime scene nor does it appear in the DNA database. Profile E's DNA is found at the scene.

4. Each individual whale has its own DNA profile. By profiling the tissue it is possible to identify how many whales were killed by simply counting the number of different profiles. It can then be determined if the whale meat sold was from one whale or many.

Key Terms: Word Find (page 156)

```
A M P L I F I C A T I O N F E O Y Z X N B C E K Q
N I P C X N U C L E O T I D E A A W M I T Z T K B
P J B Y P Y P V E X M C O J L V F U E R R O C I F
Z G A Y C L R M T J I C I X E J A E S S A W P B V
Y D Y W Y V O N S M C U T D C O N W S C N A E I P
E T N I Y S T S Z I R R Q T X F K E Y S D S B B
E G A A O J E U T R O Q U Z R Y T L N F L G W R A
X P B B P G I E K G S P I A O R R I G N A K A G Q
P X I F K R N T J F A T E M P L A T E S T R A N D
R R P Y Z E O H P U T J O L H N N X R N I B I Z M
E L L C G U B F L V E O S Z O B S C R M O F Y U T
S S V Z R H G D I V L A A X R A F A N R N R M N W
S U C D N Q W N D L L C E Y E N E S A X Q O E H Y
I N T R O N L X W R I K J B S T R X N R C H Q H Z
O J T D N A T U F V T N L Q I F R K T R I P L E T
N F N W R P S O L H E C G A S Z N B L T B G I D T
J T R A N S C R I P T I O N K Z A D R H V N S L Z
```

Characteristics of Life (page 158)

1. Three of: cytoplasm (nutrient "soup"), plasma membrane, genetic material, metabolism.

2. (a) **Size**: Viruses are very small: generally 50-500 times smaller than a typical prokaryotic cell and up to 5000 times smaller than a eukaryote cell.
   (b) **Metabolism**: Cells have metabolic activity; there are chemical reactions taking place much of the time. A virus has no cytoplasm and no metabolism of its own. It relies on the metabolism of its host cell.
   (c) **Organelles**: Viruses have no organelles unlike cells, most of which have organelles which carry out specific roles in the cell.
   (d) **Genetic material**: Viruses have a single or double stranded chromosome which can be RNA or DNA. Cells have only double stranded DNA chromosomes. In eukaryotes the chromosomes are contained within a nuclear membrane.
   (e) **Life cycle**: Outside a living cell viruses exist as inert particles, adopting a "living" programme only when they invade a host cell and can take over the cellular machinery of the cell. At times, they may integrate into the host cell's chromosome and remain latent. Cells are generally either "alive" (when there is metabolic activity) or dead (no metabolic activity).
   Note: There are exceptions to this generalisation, e.g. bacterial endospores, which are special resting stages with no metabolic activity.

3. Multicellular organisms are said to show emergent properties because they have properties that go

beyond those of a single cell or unicellular organism, e.g. specialised cell types with differing functions.

Bacterial Cells (page 159)

1. (a) The nuclear material (DNA) is not contained within a defined nucleus with a nuclear membrane.
   (b) Membrane-bound cellular organelles (e.g. mitochondria, endoplasmic reticulum) are missing.
   (c) Single, circular chromosome sometimes with accessory chromosomes called plasmids.

2. (a) Locomotion - flagella enable bacterial movement.
   (b) Fimbriae are shorter, straighter, and thinner than flagella. They are used for attachment, not locomotion.

3. The bacterial cell wall lies outside the plasma (cell surface) membrane. It is a semi-rigid structure composed of a macromolecule called peptidoglycan, and contains varying amounts of lipopolysaccharides and lipoproteins.

The Structure of Viruses (160)

1. Virion structure: a single type of nucleic acid (DNA or RNA, double or single stranded) and a few enzymes (e.g. reverse transcriptase in HIV), enclosed in a protein coat or capsid. The capsid may be covered by an envelope of lipid, protein, or carbohydrate.

2. Viruses do not conform to any of the criteria by which other life-forms are conventionally classified.

3. They are entirely dependent on using the host's cellular machinery to reproduce.

4. Viruses require living cells in which to replicate and they are very specific to their host. Successful culture requires culture of the right type of cell, and often the cell itself must be in the appropriate physiological state.

Replication in Animal Viruses (page 161)

1. Glycoprotein spikes are important in host recognition and attachment of the virus to the host cell.

2. (a) Endocytosis is the means by which foreign material is normally engulfed by cells, prior to being destroyed. This response of the cell enables the virus to gain entry into the cell.
   (b) Viral DNA replicated in the host cell's nucleus.
   (c) Viral proteins synthesised in the host cell's cytoplasm.

3. (a) HIV enters a cells by attaching to the CD4 receptors on a T cell, and fusing with the cell's plasma membrane.
   (b) Reverse transcriptase transcribes the viral RNA into viral DNA. This must occur for the viral genes to be able to integrate into the host's chromosomes where it stays as a provirus.
   (c) The provirus remains integrated with the host chromosome and persists as a latent infection. This means that it can reinfect new host cells whenever the DNA is replicated.

4. (a) Attachment: The virion comes into contact with a cell and adheres to receptor sites on the cell surface. The attachment structures on the viral surface match the receptors on the host cell.
   (b) Penetration: The host cell engulfs the attached viral particle by endocytosis.
   (c) Uncoating: The host's enzymes degrade the protein coat and release the viral nucleic acid into the cell.
   (d) Biosynthesis: The synthesis (assembly) of new infective virions.
   (e) Release: The active virions are budded off from the host cell by exocytosis.

Under Siege (page 163)

1. A parasite or pathogen benefits from entering a host's body because it offers a regular, easily accessible source of food, and a stable, relatively constant environment in which the parasite or pathogen can complete part or all of its life cycle.

2. (a) and (b) any order: Washing hands after using the toilet or blowing nose. Sneezing or coughing into a disposable tissue, rather than into a hand or into the air.

3. (a) **Transmission**: Animal vector (mosquito); this is an example of biological transmission where the vector is part of the pathogen's life cycle.
   Portal of entry: Through the skin (via a bite).
   (b) **Transmission**: Indirect contact transmission (e.g. as when injured with an infected rose thorn).
   Portal of entry: Through the skin (via a puncture).
   (c) **Transmission**: Waterborne vehicle transmission.
   Portal of entry: Gastrointestinal tract.
   (d) **Transmission**: Direct contact transmission and droplet transmission.
   Portal of entry: Respiratory tract.
   (e) **Transmission**: Droplet transmission.
   Portal of entry: Respiratory tract.
   (f) **Transmission**: Direct (sexual contact) and indirect (e.g. syringes) contact transmission, vehicle transmission through transfusions of infected blood.
   Portal of entry: Urinogenital openings, gastrointestinal tract.
   (g) **Transmission**: Direct (sexual) contact transmission.
   Portal of entry: Urinogenital openings.

The First Line of Defence (page 164)

1. The natural population of (normally non-pathogenic) microbes can benefit the host by preventing overgrowth of pathogens (through competitive exclusion).

2. The skin provides a physical barrier to prevent pathogens entering the body. Skin secretions (serum and sweat) contain antimicrobial chemicals which inhibit microbial growth.

3. A non-specific response acts against any foreign body or pathogen. The response is immediate and always at the same response level (a maximum response). There is no capacity for immunological learning.

The Body's Defences (page 165)

1. The **first line of defence** provides non-specific resistance by forming a physical barrier to the entry of pathogens. Chemical secretions from the skin, tears, and saliva also provide antimicrobial activity and help destroy pathogens and wash them away. The **second line of defence** provides non-specific resistance operating inside the body to inhibit or destroy

pathogens (irrespective of what type of pathogen is involved). Whereas the **third line of defence** provides specific defence resistance against particular pathogens once they have been identified by the immune system (antibody production and cell-mediated immunity).

2. **Specific** resistance refers to defence against particular (identified) pathogens. It involves a range of specific responses to the pathogen concerned (antibody production and cell-mediated immunity). In contrast, **non-specific** resistance refers to defence against any type of pathogen.

3. Functional role for (a)–(c) as follows:
   (a) Destroys pathogens directly by engulfing them.
   (b) Produced against specific pathogens, antibodies bind and destroy pathogens or their toxins.
   (c) Some secretions (sebum) have a pH unfavourable to microbial growth. The pH of gastric juice is low enough to kill microbes directly. Other secretions (tears, saliva) wash microbes away, preventing them settling on surfaces, sweat contains an enzyme that destroys some types of bacterial cell walls, urine flushes potential pathogens from the urinary tract.

4. The **major histocompatibility complex (MHC)** is a cluster of tightly linked genes on chromosome 6 in humans. The genes code for MHC antigens (proteins) that are attached to the surfaces of all body cells and are used by the immune system to distinguish its own tissue from that which is foreign.

5. The self-recognition system allows the body to immediately identify foreign tissue e.g. a pathogen, and mount an immune attack against it for the protection of the body's own tissues.

6. Self-recognition is undesirable (any of):
   – **During pregnancy** [some features of the self-recognition system are disabled to enable growth (to full term) of what is essentially a large foreign body].
   – **During tissue and organ grafts/transplants** from another human (allografting) or a non-human animal (xenografting). [Such grafts are usually for the purpose of replacing rather than repairing tissue (e.g. grafting to replace damaged heart valves). For these grafts, tissue-typing provides the closest match possible between recipient and donor. The self-recognition system must also be suppressed indefinitely by immunosuppressant drugs].

The Action of Phagocytes (page 167)

1. Neutrophils, eosinophils, macrophages.

2. By looking at the ratio of white blood cells to red blood cells (not involved in the immune response). An elevated white blood cell count (specifically a high neutrophil count) indicates microbial infection.

3. Microbes may be able to produce toxins that kill phagocytes directly. Others can enter the phagocytes, completely filling them and preventing them functioning or remaining dormant and resuming activity later.

Inflammation (page 168)

1. (a) Increased diameter and permeability of blood vessels. **Role**: Increases blood flow and delivery of leukocytes to the area. Aids removal of destroyed microbes or their toxins. Allows defensive substances to leak into the tissue spaces.
   (b) Phagocyte migration and phagocytosis. **Role**: To directly attack and destroy invading microbes and foreign substances.
   (c) Tissue repair. **Role**: Replaces damaged cells and tissues, restoring the integrity of the area.

2. Ability to squeeze through capillary walls (amoeboid movement). Ability to engulf material by phagocytosis.

3. Histamines and prostaglandins attract phagocytes to the site of infection.

4. Pus is the accumulated debris of infection (dead phagocytes, damaged tissue, and fluid). It accumulates at the site of infection where the defence process is most active.

Fever (page 169)

1. The high body temperature associated with fever intensifies the action of interferon (a potent antiviral substance). Fever also increases metabolism, which is associated with increased blood flow. These changes increase the rate at which white blood cells are delivered to the site of infection and help to speed up the repair of tissues. The release of interleukin-1 during fever helps to increase the production of T cell lymphocytes and speeds up the immune response.

2. **1:** Macrophage ingests a microbe and destroys it.
   2: The release of endotoxins from the microbe induces the macrophage to produce interleukin-1 which is released into the blood.
   3: Interleukin-1 travels in the blood to the hypothalamus of the brain where it stimulates the production of large amounts of prostaglandins.
   4: Prostaglandins cause resetting of the thermostat to a higher temperature, causing fever.

The Lymphatic System (page 170)

1. **Lymph** has a similar composition to tissue fluid but has more leukocytes (derived from lymphoid tissues). **Note**: Tissue fluid is similar in composition to plasma (i.e. containing water, ions, urea, proteins, glucose etc.) but lacks the large proteins found in plasma.

2. Lymph returns tissue fluid to general circulation, and with the blood, circulates lymphocytes around the body.

3. (a) **Lymph nodes**: Filter foreign material from the lymph by trapping it in fibres. They also produce lymphocytes.
   (b) **Bone marrow**: Produce many kinds of white blood cells: monocytes, macrophages, neutrophils, eosinophils, basophils, T and B lymphocytes.

The Immune System (page 171)

1. (a) **Humoral immune system**: Production of antibodies against specific antigens. The antibodies disable circulating antigens.
   (b) **Cell-mediated immune system**: Involves the production of T cells which destroy pathogens or their toxins by direct contact or by producing substances that regulate the activity of other cells in

the immune system.

2. In the bone marrow (adults) or liver (foetuses).

3. (a) Bone marrow   (b) Thymus

4. (a) **Memory cells**: Retain an antigen memory. They can rapidly differentiate into antibody- producing plasma cells if they encounter the same antigen again.
   (b) **Plasma cells**: Secrete antibodies against antigens (very rapid rate of antibody production).
   (c) **Helper T cells**: Activate cytotoxic T cells and other helper T cells. Also needed for B cell activation.
   (d) **Suppressor T cells**: Regulate the immune system response by turning it off when antigens disappear.
   (e) **Delayed hypersensitivity T cells**: Cause inflammation in allergic responses and are responsible for rejection of transplanted tissue.
   (f) **Cytotoxic T cells**: Destroy target cells on contact (by binding and lysing cells).

5. **Immunological memory**: The result of the differentiation of B cells after the first exposure to an antigen. Those B cells that differentiate into long lived memory cells are present to react quickly and vigorously in the event of a second infection.

Antibodies (page 173)

1. **Antibodies** are proteins produced in response to antigens; they recognise and bind antigens. **Antigens** are foreign substances (often proteins) that promote the formation of antibodies (invoke an immune response).

2. (a) The immune system must be able to recognise self from non-self so that it can recognise foreign material (and destroy it) and its own tissue (and not destroy it).
   (b) During development, any B cells that react to the body's own antigens are selectively destroyed. This process leads to self tolerance.
   (c) Autoimmune disease (disorder).
   (d) Any two of: Grave's disease (thyroid enlargement), rheumatoid arthritis (primarily joint inflammation), insulin-dependent diabetes mellitus (caused by immune destruction of the insulin-secreting cells in the pancreas), haemolytic anaemia (premature destruction of red blood cells), and probably multiple sclerosis (destruction of myelin around nerves).

3. Antibodies inactivate pathogens in four main ways: **Neutralisation** describes the way in which antibodies bind to viral binding sites and bacterial toxins and stop their activity. Antibodies may also **inactivate particulate antigens**, such as bacteria, by sticking them together in clumps. Soluble antigens may be bound by antibodies and fall out of solution (**precipitation**) so that they lose activity. Antibodies also activate **complement** (a defence system involving serum proteins), tagging foreign cells so that they can be recognised and destroyed.

4. (a) **Phagocytosis**: Antibodies promote the formation of inactive clumps of foreign material that can easily be engulfed and destroyed by a phagocytic cell.
   (b) **Inflammation**: Antibodies are involved in activation of complement (the defence system involving serum proteins which participate in the inflammatory response and other immune system activities).
   (c) **Bacterial cell lysis**: Antibodies are involved in

tagging foreign cells for destruction and in the activation of complement (the defence system involving serum proteins which participate in the lysis of foreign cells).

Acquired Immunity (page 175)

1. (a) **Active immunity** is immunity that develops after the body has been exposed to a microbe or its toxins and an immune response has been invoked.
   (b) **Naturally acquired** active immunity arises as a result of exposure to an antigen such as a pathogen, e.g. natural immunity to chickenpox. **Artificially acquired** active immunity arises as a result of vaccination, e.g. any childhood disease for which vaccinations are given: diphtheria, measles, mumps, polio etc.

2. (a) **Passive immunity** describes the immunity that develops after antibodies are transferred from one person to another. In this case, the recipient does not make the antibodies themselves.
   (b) **Naturally acquired** passive immunity arises as a result of antibodies passing from the mother to the foetus/infant via the placenta/breast milk. **Artificially acquired** passive immunity arises as a result of injection with immune serums e.g. in antivenoms.

3. (a) Newborns need to be supplied with maternal antibodies because they have not yet had exposure to the everyday microbes in their environment and must be born with operational defence mechanisms.
   (b) The antibody "supply" is (ideally) supplemented with antibodies in breast milk because it takes time for the infant's immune system to become fully functional. During this time, the supply of antibodies received during pregnancy will decline.

Vaccination (page 176)

1. (a) 280 days
   (b) Antibody levels gradually build to a small peak after 40 days, then gradually decline to very low levels.
   (c) Antibody levels rise very rapidly to a peak (much higher than that achieved after the first injection). Levels then decline slowly over a long period of time.
   (d) The immune system has been "primed" or prepared to respond to the antigen by the first exposure to it (this initial response took a considerable time). When the cells of the immune system receive a second exposure to the same antigen they can respond quickly with rapid production of antibodies.

Eradicating Disease (page 177)

1. High vaccination rates increase the rates of immunity within a population, so fewer people will contract the disease with each outbreak. Transmission of the disease is limited because there are fewer susceptible hosts for the disease to exploit, until eventually the disease no longer occurs in the population.

2. Several research institutions maintain the smallpox virus for ongoing research. Smallpox researchers must be vaccinated to prevent accidental contamination. More recently there have been concerns about the smallpox virus being used as a weapon in bioterrorism.

4. The number of reported cases of whooping cough

increased significantly. They only began to drop again when vaccination rates increased again.

Bacterial Diseases (page 178)

1. Students should include reference to the following:
- The use of toxins to break down the host cells or connective tissue. This allows the bacteria to spread throughout the body.
- The bacteria release enzymes to break down the clotting protein fibrin. This allows the bacteria to disperse readily through the host's blood system.
- The bacteria release cytotoxic substances that destroy the host's phagocytic cells.
- Fimbriae on the bacterial surface allow them to attach to mucous membranes and attack the host's tissues.

2. Airborne diseases are very easily spread when the infected person coughs or sneezes. Immunisation (vaccination) against airborne bacterial diseases would prevent a person becoming infected even if they did come in contact with a contagious person and help reduce the spread of the disease.

Tuberculosis (page 179)

1. Both forms of TB are caused by the same bacterium, *Mycobacterium tuberculosis*. In the case of a latent TB infection, the bacteria are present in the body (alive but inactive) but the disease is dormant and the person has no symtoms and is not infectious. Active TB usually develops in people under stress (e.g. malnourished, immune compromised, living in overcrowded situations). In people with active TB, the bacteria are growing and the patients show symptoms (most often associated with pulmonary TB), and are highly contagious. If left untreated they can die.

2. The body's defence system can isolate the bacteria in nodules (tubercles). In this dormant state, the bacteria do not have the capacity to cause disease symptoms and the person is not infectious.

3. (a) Inhaling the TB pathogen results in a defence response from lung macrophages, causing nodules to form in the lung tissue. These regions can rupture through the alveolar walls to release bacteria into the airways and they are replaced by scarring and areas of dying tissue. Fluid accumulates, causing congestion, and lung function is impaired.
   (b) TB is transmitted by airborne droplets, which are coughed or sneezed out following the rupture of tubercles, which releases bacilli into the airways.

4. Africa and Asia contain some of the poorest and most underprivileged people in the world. Overcrowding, poverty, and high disease rates are characteristic of these regions. These regions also have the highest rates of HIV/AIDS, so many people are immune-compromised. All of these factors are known to contribute to the likelihood of TB infection.

5. Previously infected people are susceptible to developing TB again even after successful treatment. In addition, TB patients often do not complete the course of medication and so never get rid of the infection completely. This means it can reoccur, particularly in high risk groups, e.g. alcoholics, people with HIV/AIDS.

HIV and AIDS (page 181)

1. HIV attacks the body's immune system, leaving the body vulnerable to infection from microbes that could not normally infect a healthy person.

2. (a) The virus rapidly increases in numbers within the first year of infection, followed by a large drop off in numbers in the second year. Over the next 3-10 years, the HIV population gradually increases again.
   (b) The helper T cell numbers respond to the initial infection by increasing in numbers. After about a year, their numbers steadily decrease as they are attacked and destroyed by the HIV.

3. Any three of: Blood or blood products, vaginal secretions, breast milk, across the placenta, shared needles among intravenous drug users, sexual intercourse: both homo- and heterosexual.

4. **HIV positive**: Blood tests have detected the presence of HIV in blood samples from a person (even though they may not have exhibited any symptoms).

5. Blood donated by the public and used to obtain a blood clotting factor (Factor VIII) for haemophiliacs was contaminated with HIV from donors already infected with the virus. This is particularly the case in countries where people are paid to donate blood.

6. HIV mutates very rapidly, changing its protein coat. Any vaccine is quickly rendered useless as the virus has already changed. There are also many different strains of HIV, each requiring a different vaccine.

7. Such people have a genetic makeup that predisposes their metabolism to HIV resistance - the HIV virus does not replicate effectively inside their bodies. By determining how these people are different, it may be possible to develop drugs that mimic the effect.

Epidemiology of AIDS (page 183)

1. Men are likely to have **multiple partners concurrently**, and **avoid condom use**. **Women have little power** or say in controlling their circumstances and are at risk from partners engaging in risky sexual practices. The lack of employment opportunities in rural areas results in an **itinerant population of men**, moving between cities for work and developing sexual networks involving high risk behaviours. They contract HIV and return to spread it through their rural communities.

2. (a) Age structure: A high proportion of individuals of reproductive age are infected (and die) so that the population becomes dominated by nonreproductive individuals (old people and children). **Note**: many infants, infected *in utero* or during delivery, also die).
   (b) Local economies will suffer through the removal (or incapacitation) of the sector of the population that normally provides most of the income. This group normally also supports children and elderly, so the economy also bears the cost of caring for the HIV infected, their children, and their old people.

3. Sub-Saharan Africa countries are underdeveloped in their technologies and economies and do not have the resources to fund comprehensive AIDS treatment programme. Economically developed countries, which could support such drug programmes, choose not to.

4. HIV-1: Recombination of two existing strains of SIV in a subspecies of common chimpanzee to produce a new strain. There was then cross species transmission of the virus (zoonosis), probably through the handling and consumption of infected chimpanzee carcasses.
HIV-2: Mutation of an existing SIV found in the sooty mangabey (SIV sm). Transmission as above.

Antibiotics (page 185)

1. Ideally, an antimicrobial drug should have selective toxicity, targeting and killing the pathogen without harming the host. There is a wide range of antibiotics available; broad spectrum antibiotics, effective against a wide range of bacteria, are useful when the identity of the pathogen is unknown and a treatment decision must be made quickly. Narrow spectrum antibiotics are useful when the pathogen is known and can be targeted directly. The latter are the preferred choice as they limit the disturbance to the body's own microbial flora. Some patients exhibit side effects, ranging from discomfort to anaphylaxis, but the vast majority of people experience few difficulties with their use.

2. (a) **Antibiotic resistance** refers to the resistance some bacteria develop to antibiotics that would normally inhibit their growth or kill them. In other words, they no longer show a reduction in growth response in the presence of the antibiotic.
   (b) It is important to finish a course of antibiotics so that the chances of survival of resistant mutants are minimised. If the course ends too quickly, more resistant cells may survive and flourish when antibiotic levels in the blood fall.

3. (a) **Bacteriostatic antibiotics** inhibit the growth of bacteria by interfering with protein production, DNA replication, or other aspects of their metabolism. They do not kill the bacteria. **Bacteriocidal antibiotics** kill the bacterial cells outright. The mode of action is often to inhibit the formation of the cell wall, to break the cell wall, or damage the plasma membrane.
   (b) **Bacteriostatic:** Chloramphenicol is a bacteriostatic antibiotic to *S. pneumoniae*. Bacterial numbers remain constant, there is no growth or reduction in numbers.
   Bacteriocidal: Ampicillin is bacteriocidal to *S. pneumoniae*. Bacterial numbers are rapidly reduced after exposure to this antibiotic.

4. (a) Antibiotic A is the most effective antibiotic because it produced the largest zone of clearance.
   (b) Disc 3 (5.5 µg ml^{-1}) and disc 4 (7 µg ml^{-1}) both had the same level of activity (they produced the same zone of clearance). However, Disc 3 is the most effective because it provides the same response as disc 4 but at a lower concentration.

Resistance in Pathogens (page 187)

1. Factors contributing to the rapid spread of drug resistance in pathogens include:
   - The typically high mutation rates in viruses, bacteria, and protozoa, combined with short generation times (rapid generational turnover). Note: In bacteria and viruses, high mutation rates are the result of a higher error rate during DNA replication than is typical in most eukaryotic genomes.
   - Strong selection pressures imposed by drug use, coupled with misuse of drugs (e.g. using antibiotics against viral infections), poor patient compliance (patients not taking drugs as prescribed), and the poor quality of available drugs (especially to impoverished populations).

2. Answer will depend on student choice. All mechanisms are based on mutations that confer greater fitness in the prevailing environment:
   - For bacteria: Genes for drug resistance (often carried on plasmid DNA) arise through mutation. In an environment that selects against susceptible strains, resistant bacteria will survive and increase in numbers. Genes for drug resistance are also easily transferred between strains, leading to a spread in resistance while that selection pressure remains. Drug resistance in bacteria can be conferred via a number of mechanisms, including inactivation of the drug, alteration of the drug's target, or alteration in the permeability of the bacterial cell to the drug.
   - For HIV: Resistance may arise though a single mutation or through the accumulation of specific mutations over time. Such mutations may alter the binding capacity of the drug or susceptibility of the virus to the drug. Alternatively, resistance may arise as a result of naturally occurring polymorphisms, which gain favour in the selective environment.
   - Chloroquine resistance in *Plasmodium falciparum* is based on the fact that they accumulate significantly less chloroquine than susceptible parasites. The mechanism for this appears to be due to a mutation conferring an enhanced ability to release the chloroquine from the vesicles in which it normally accumulates in the cell.

3. Health authorities can target multiple drug resistance by administering several drugs at the same time (what is called a drug cocktail). This does two things: it targets different stages of the pathogen's life cycle and it minimises the number of pathogens in the body. This multi-pronged attack reduces the pathogen "load" and minimises its reproduction/replication rate, while also reducing the chance of further mutations arising.

The Basis of Resistance (page 188)

1. Horizontal gene transmission describes the transfer of genetic material directly between bacteria by conjugation, transduction, or transformation. Vertical gene transmission describes the passing of genetic information from generation to generation by cell division.

2. When bacteria acquire several different mechanisms of resistance there is a much greater chance that they will become resistant to different classes of antibiotics. If this occurs, the treatment options against that particular pathogen are greatly reduced.

Avoiding Detection (page 189)

1. HIV avoids the immune system by rapidly mutating (around 1 mutation per replication), and produces more HIV variants than the immune system can produce antibodies for. It attacks T helpers cells which regulate the immune response. HIV has antibody binding sites within deep valleys where the body's antibodies cannot bind to them. TB bacteria inhibit macrophage action

and live inside the cells. A thick waxy coat protects them from the immune system.

2. The human immune system is constantly adapting to recognise and remove pathogens. However, the pathogens are also evolving mechanisms to avoid detection by the immune system, so the advantages gained by the human host are quickly negated by changes in the pathogen.

Hospital Acquired Infection (page 190)

1. Hospitals contain a large number of infectious agents simply because of the number of patients and visitors that pass through the hospital. The transfer of these agents to patients with weakened immune systems has the potential to cause serious infection and lengthen a patient's stay, or result in their death. Strict hygiene procedures are designed to break the chain of infection and reduce pathogen numbers. Procedures include hand washing, containment of patients with drug resistant microbes, and washing and sterilisation of hospital linen, clothing, and surgical equipment.

Time of Death... (page 191)

1. Establishing the time of death is particularly important in homicide cases. It can help investigators to identify suspects and eventually the murderer. The whereabouts of suspects leading up to, and at, the estimated time of death allows investigators to include or dismiss them in the investigation by eliminating people with verified alibis. Establishing time of death also allows investigators to determine the victim's final movements and the movements of any people connected to the victim. Establishing time of death can also be useful for coroners investigating cases where people have died of unknown or suspicious causes.

2. (a) Rigor mortis sets in around 3 hours after death and reaches a maximum after around 18 hours before weakening and disappearing around 36 hours. Its presence or absence allows estimation of time of death to within hours if the body is detected during the first 36 hours.
   (b) Livor mortis is evident after 2 hours and reaches its maximum appearance between 8 and 12 hours. The distribution of livor mortis also allows the position of the body at death to be determined. It can be used to determine if the body has been moved after death.
   (c) The change in core (rectal) temperature can be used to calculate the time since death using Glaisters equation. Temperature declines at a linear rate, but it depends on the environmental temperature and the health of the deceased. These factors must be taken into account when using temperature to determine time of death.

3. Forensic entomology provides another way to estimate time of death. Blow flies are the most commonly found insect on a body, and are usually the first insect to colonise a dead body. The blow fly life cycle last around two weeks from egg to pupae hatching, so the developmental stage of the larvae can estimate time of death to within a day (sometimes hours if found early). The number and species of larvae present helps determine time of death in more decomposed bodies. In addition, body lice will leave a body within 2-3 days.

4. Signs of rigor mortis indicates death was between 3 and 36 hours ago. The presence of first stage maggot larvae indicate death occurred between 12-24 hours ago. Glaiser temperature equation estimates death occurred about 15 hours ago ((37-24) x 1.2 = 15.6). Livor mortis indicates at least 8 hours since death. Time of death is probably around 15 hours, definitely within 24. The death is suspicious because the body has been moved to a face down position but livor mortis is present on the victims back (indicating they were in this position at death and have been subsequently moved).

Muscle Structure and Function (page 197)

1. Smooth muscle, also called involuntary muscle, has spindle shaped cells with one central nucleus per cell and a smooth appearance with no striations. Its contractions are diffuse and it is not under conscious control so it is involved in movement of visceral organs such as the gut. Striated (skeletal) muscle is voluntary and is responsible for the skeletal muscle movement over which we have conscious control. It has a striated or striped appearance and the cells are multinucleate with peripheral nuclei. Cardiac muscle is involuntary muscle responsible for the contraction of the heart. Although it is striated, it does not fatigue in the same way as skeletal muscle. There is one nucleus per cell and structures called intercalated discs (electrical junctions) join individual cells.

2. (a) The banding pattern results from the overlap pattern of the thick and thin filaments (dark = thick and thin filaments overlapping, light = no overlap).
   (b) I band: Becomes narrower as more filaments overlap and the area of non-overlap decreases. H zone: Disappears as the overlap becomes maximal (no region of only thick filaments). Sarcomere: Shortens progressively as the overlap becomes maximal.

3. Fast twitch fibres are large, pale fibres with rapid contraction rates, rapid rates of ATP production, and high power generation. They work anaerobically, but fatigue quickly, and these properties suit them to short bursts of activity where maximal force is required quickly (as in sprinting or power lifting). Slow twitch fibres, in contrast, are darker (red) due to the presence of large amounts of myoglobin, they are small in diameter and have a slower rate of ATP production and a slower contraction rate. They work aerobically, but fatigue slowly, so they are suited to activities where endurance and prolonged activity at a sustainable level is required.

The Mechanics of Movement (page 199)

1. (a) Prime mover: the muscle primarily responsible for the movement.
   (b) Antagonist: the muscle that opposes the prime mover. i.e. relaxes when prime mover contracts. Its action can be protective in preventing over-stretching of the prime mover during contraction.
   (c) Synergist: assists the prime mover by fine-tuning the direction of limb movement.

2. Muscles can only contract and relax, therefore they can only pull on a bone; they cannot push it. To produce movement, two muscles must act as **antagonistic** pairs to move a bone to and from different positions.

3. Muscles have an origin on one (less moveable) bone and an insertion on another (more moveable) bone. When the muscle contracts across the joint connecting the two bones, the insertion moves towards the origin, therefore moving the limb. To raise a limb, the flexor (prime mover in this case) contracts pulling the limb bone up (extensor/antagonist relaxed). To lower the limb, the extensor contracts, pulling the limb down (flexor relaxed).

4. Bones are rigid and can only move at joints where they come together. The degree of movement allowed depends the type of joint. The bones of the limbs, which need to be freely moveable are connected by synovial joints. Body parts where movement is not desirable, such as the joints of the skull, are largely rigid.

5. (a) Radius (radial tuberosity)
   (b) Ulna
   (c) Triceps
   (d) Rotation of ulna and radius

6. (a) Elbow extension
   (b) Triceps

7. (a) Nodding head = flexion and extension
   (b) Hitching a ride = abduction

The Sliding Filament Model (page 201)

1. (a) Myosin: Has a moveable head that provides a power stroke when activated.
   (b) Actin: Two protein molecules twisted in a double helix shape that form the thin filament of a myofibril.
   (c) Calcium ions: Bind to the blocking molecules, causing them to move and expose the myosin binding site.
   (d) Troponin-tropomyosin: Bind to actin molecule in a way that prevents myosin head from forming a cross bridge.
   (e) ATP: Supplies energy for flexing of the myosin head (power stroke).

2. (a) By changing the frequency of stimulation, so that fibres receive impulses at a greater rate (frequency summation).
   (b) By changing the number and size of motor units recruited (a few motor units = a small contraction, maximum number of motor units = maximum contraction).

3. (a) Calcium ions and ATP.
   (b) Calcium ions are released from a store in the sarcoplasmic reticulum when an action potential arrives. ATP is present in the muscle fibre and is hydrolysed by ATPase enzymes on the myosin.

Cellular Respiration (page 202)

1. (a) Glycolysis: cytoplasm
   (b) Krebs cycle: matrix of mitochondria
   (c) Electron transport chain: cristae (inner membrane surface) of mitochondria.

2. (a) Substrate-level phosphorylation: Formation of ATP by the transfer a phosphate group from a substrate (e.g. a phosphorylated 6C sugar) to ADP directly, with no involvement of an electron transport chain.
   (b) Oxidative phosphorylation: The process by which glucose is oxidised in a series of redox reactions and the energy released in the electron transfers is coupled to ATP synthesis.

The Biochemistry of Respiration (page 203)

1. (a) 6 carbon atoms
   (b) 3 carbon atoms (glucose split into two)
   (c) 2 carbon atoms (1 carbon lost as CO_2)
   (d) 6 carbon atoms (2-carbon acetyl added to 4 carbon)
   (e) 5 carbon atoms (1 carbon lost as CO_2)
   (f) 4 carbon atoms (1 carbon lost as CO_2)

2. (a) Glycolysis: 2 ATPs
   (b) Krebs cycle: 2 ATPs
   (c) Electron transport chain: 34 ATPs
   (d) Total produced: 38 ATPs

3. Carbon atoms are released as carbon dioxide (CO_2) gas and breathed out through gas exchange surfaces.

4. (a) Hydrogen atoms supply energy in the form of high energy electrons. **Note**: These are passed along the respiratory chain, losing energy as they go. The energy released is used to generate ATP.
   (b) Oxygen is the final electron acceptor at the end of the respiratory chain.

5. (a) ATP is generated by **chemiosmosis**.
   (b) **In brief**: The synthesis of ATP is coupled to electron transport and movement of hydrogen ions.
   In more detail: Energy from the passage of electrons along the chain of electron carriers is used to pump protons (H^+), against their concentration gradient, into the intermembrane space, creating a high concentration of protons there. The protons return across the membrane down a concentration gradient via the enzyme complex, ATP synthetase (also called ATP synthase or ATPase), which synthesises the ATP.

Anaerobic Pathways (page 205)

1. **Aerobic respiration** requires the presence of oxygen and produces a lot of useable energy (ATP). **Fermentation** does not require oxygen and uses an alternative H^+ acceptor. There is little useable energy produced (the only ATP generated is via glycolysis).

2. (a) $2 \div 38 \times 100 = 5.3\%$ efficiency
   (b) Only a small amount of the energy of a glucose molecule is released in anaerobic respiration. The remainder stays locked up in the molecule.

3. The build up of toxic products (ethanol or lactate) inhibits further metabolic activity.

Measuring Respiration (page 206)

1. (a) RQ at 20°C/48 h = $1.97 \div 2.82 = $ **0.7** (0.698)
   (b) RQ at 20°C/1 h = $2.82 \div 2.82 = 1$. After 2 days without feeding the cricket was metabolising fats for energy. Shortly (1 hour) after feeding it was metabolising only carbohydrate (RQ = 1).

2. (a) RQ of two seedlings during early germination:

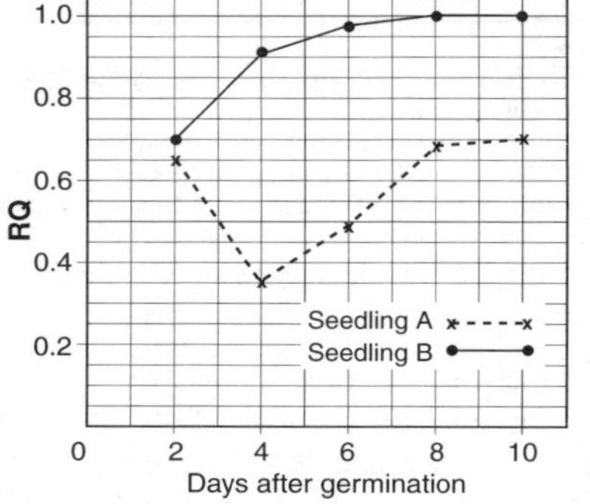

(b) Both seedlings began their germination metabolising primarily fats for energy. However, while seedling A continued to metabolise mainly fats (with some synthesis of carbohydrate and organic acids), throughout the 10 day period, seedling B rapidly moved to metabolising carbohydrate alone. **Note**: The value of 0.91 (seedling B) may have been the result of protein metabolism alone or (more likely) respiration of a mix of fat and glucose.

Lactate and Oxygen Debt (page 207)

1. Lactic acid quickly builds up as a waste product of the glycolytic pathway. Lactic acid inhibits glycogen breakdown and impedes muscular contraction, so the time period for its use is limited.

2. (a) Shaded region **A**: Oxygen deficit: the amount of oxygen needed for aerobic supply but not supplied by breathing i.e. an oxygen deficit builds up.
   (b) Shaded region **B**: Oxygen debt: the extra oxygen required (taken in) despite the drop in energy demand. The debt is used to replace oxygen reserves, restore CP, and break down lactic acid. **Note**: Both components (deficit and debt) are often used synonymously (as oxygen debt) although they are not quite the same. The deficit is the oxygen shortfall incurred; the debt is the amount of oxygen required to restore oxygen and energy stores to resting levels. They may not be the same.

Key Terms: Crossword (page 208)

Answers Across	Answers Down
1. Myofibril	2. Fast twitch
4. ATPase	3. Flexor
5. Chemiosmosis	6. Sliding filament
7. Ligament	9. Extensor
8. Slow twitch	10. Troponin
12. Tropomyosin	11. Aerobic
13. Krebs cycle	14. Muscle
15. Anaerobic	
16. Glycolysis	
17. Actin	
18. Coenzyme	
19. Cellular respiration	
20. Tendon	

Principles of Homeostasis (page 210)

1. **Receptors** (detect stimuli), **control and coordination centre** (integration of signals and coordination of response), **effectors** (implement appropriate response).

2. Negative feedback mechanisms enable maintenance of a steady state internal environment despite fluctuations in the external environment (e.g. rising air temperature). Negative feedback mechanisms are self-correcting so that physiological systems are stabilised against excessive change.

Maintaining Homeostasis (page 211)

1. Two mechanisms operating to restore homeostasis after infection (a and b any two of):
   - Immune system response with the production of antibodies against the antigens of the pathogen (humoral response).
   - Immune system response with the production of T cells which recognise the antigens of the pathogen and destroy them directly (cell-mediated response).
   - Local inflammatory response (redness, pain, swelling, heat) at the site of infection.
   - Fever (widespread increase in body temperature).
   - The production of antimicrobial substances such as interferon and interleukin-1.
   - Phagocytosis of pathogen by white blood cells.
   All the above aim to destroy the pathogen and/or its toxins and assist a return to homeostasis.

2. Two mechanisms by which responses to stimuli are brought about and coordinated (a and b in any order):
   (a) **Hormonal response** to stimuli: endocrine glands respond to a stimulus (e.g. a nerve impulse or another hormone or metabolite) by producing hormones which bring about an appropriate physiological response.
   (b) **Nervous response** to stimuli: direct stimulation of nerves from a sensory receptor causes a reaction to the stimulus. This may be a response requiring interpretation of the message by the brain or it may be a reflex.

3. Two ways in which water and ion balance are maintained, and the organs and hormones involved:
   (a) Water and ions are taken in with food and drink, helping to replace that lost through urine, faeces, and sweat. The digestive organs and digestive hormones are all involved in digestion and absorption processes.
   (b) The kidneys are the primary regulator of fluid and ions. When large quantities of fluid must be excreted, the kidney produces large amounts of dilute urine. When water must be conserved, small amounts of concentrated urine are produced. ADH (antidiuretic hormone) causes more water to be reabsorbed from the kidney (concentrating the urine). ADH secretion increases when blood volume is low. Essential ions (and glucose) are retained by active reabsorption from the kidney tubules. Another hormone, aldosterone from the adrenal glands, increases the absorption of sodium ions.

4. Two ways in which the body regulates its respiratory gases during exercise:
   (a) Increasing the rate of breathing. This increases both the rate of oxygen entering the lungs and the rate at which CO_2 leaves. It also increases the rate of loading and unloading of oxygen and CO_2 into and

out of the blood.

(b) Increasing the heart rate. This increases blood flow, which facilitates the loading and unloading of oxygen and CO_2 into and out of the blood. It also increases the speed of delivery of oxygen to working tissues (e.g. muscles) and speeds up the removal of CO_2 and other products of metabolism.

Thermoregulation in Humans (page 213)

1. Body temperature reduced by ((a) and (b) any two of):
   • Sweating (cooling by evaporation) • Reducing activity • Behavioural mechanisms such as removing clothing or seeking shade • Increasing blood flow to skin (leads to increased radiation from the skin surface) .

2. (a) **Hypothalamus**: Monitors temperature changes in the body and coordinates appropriate responses to counteract the changes.
   (b) **Skin**: Detects changes in skin temperature and relays the information to the hypothalamus. In response to input from the hypothalamus, muscles and capillaries in the skin act as effectors to bring about an appropriate thermoregulatory response.
   (c) **Nervous input to effectors** (from hypothalamus): Brings about (through stimulation of muscles) an appropriate thermoregulatory response (e.g. raising hairs, constricting blood vessels).
   (d) **Hormones**: Mediate a change in metabolic rate through their general action on body cells (adrenalin and thyroxine increase metabolic rate).

Control of Blood Glucose (page 214)

1. (a) Stimulus: Rise in the levels of glucose in the blood above a set level (about 5.5 mmol per L).
   (b) Stimulus: Fall in blood glucose levels below a set level (about 3.5 mmol per L).
   (c) Glucagon brings about the production (and subsequent release) of glucose from the liver by the breakdown of glycogen and the synthesis of glucose from amino acids.
   (d) Insulin increases glucose uptake by cells and brings about production of glycogen and fat from glucose in the liver.

2. Fluctuations in blood glucose (BG) and blood insulin levels are closely aligned. Following a meal, BG rises sharply and there is a corresponding increase in blood insulin, which promotes cellular glucose uptake and a subsequent fall in BG. This pattern is repeated after each meal, with the evening meal followed by a gradual decline in BG and insulin over the sleep (fasting) period. Negative feedback mechanisms prevent excessive fluctuations in blood glucose (BG) throughout the 24 hour period.

Homeostasis During Exercise (page 215)

1. (a) The output of the heart increases more than five times.
   (b) This increase is necessary because the heart must pump more blood in order to supply more oxygen to working muscles. In addition, an increased rate of blood flow ensures that metabolic waste products, which are produced at a higher rate during exercise, are removed as quickly as they are produced.
   (c) Heart and blood vessels (circulatory system).

2. (a) Oxygen consumption increases twenty times or more. Most of this increase occurs in the muscle.
   (b) An increase in metabolic activity associated with muscles working harder increases the requirements for oxygen (to supply the aerobic respiration of the muscle cells).
   (c) Resting muscles use about the same amount of oxygen as other tissues. During exercise, the muscles account for most of the total increase in oxygen consumption, i.e. the metabolic activity of the muscles increases greatly when they are working (contracting).

3. A trained athlete has a greater cardiac output (total blood flow) and a greater total oxygen consumption during exercise than an average man. The athlete's muscles adjust during training to working at a higher rate and during heavy exercise they demand more oxygen (and a higher blood flow) than the muscles of the average man. Note also that a trained athlete diverts slightly more blood (and oxygen) to the muscles at the expense of other tissues during exercise. This is a physiological adjustment made as a result of training to increase the working capacity of the muscles.

Cell Signalling (page 216)

1. (a) **Endocrine signalling**, where a hormone is carried in the blood between the endocrine gland/organ where it is produced to target cells.
   (b) **Paracrine signalling**, where cell signalling molecules are released to act on target cells in the immediate vicinity, e.g. at synapses or between cells during development.
   (c) **Autocrine signalling**, where cells produce and react to their own signals (e.g. growth factors from T cells stimulate the production of more T cells).

2. The three signalling types all have in common some kind of chemical messenger or signal molecule (ligand) and a receptor molecule (on the target cells, which may or may not be on the cell producing the signal).

Hormones and Exercise (page 217)

1. Many of the hormones (including testosterone, thyroxine and adrenaline) released during exercise help to raise the body's metabolic rate and allow for more energy production to meet the body's higher energy demands. For example, glucagon releases glucose into the blood so the cells have an energy source. Other hormones allow the body to keep exercising even when it is under stress. (e.g. endorphins act as pain blockers, allowing greater exertion than ordinarily possible). Growth hormone stimulates the growth and development of muscle and bone to strengthen the body, ready for the next exercise session.

2. Insulin removes blood glucose and transports it into the exercising cells to provide them with energy. Although less insulin is produced during exercise, it is more efficient at transporting blood glucose into the cells. Glucagon is also secreted during exercise, and it stimulates the liver to convert glycogen to glucose, and release it into the bloodstream. In this way blood glucose levels are regulated during exercise so the body has enough energy to carry out the exercise.

3. (a) Mechanism 1: Interaction of the hormone with

plasma membrane receptors followed by activation of a second messenger (e.g. cyclic AMP).

(b) Mechanism 2: Interaction of the hormone with intracellular receptors directly, without the involvement of a second messenger.

4. Both involve interaction of the hormone with a specific receptor molecule, which recognises the hormone and initiates a cascade of reactions leading to a physiological response.

5. The binding of a hormone molecule to a receptor, initiates a series of reactions within the cell that result in, often far reaching, effects. In many cases, hormones exert their effects by influencing the enzymes associated with membranes or genetic control.

Control of Heart Activity (page 220)

1. (a) **Sinoatrial node**: Initiates the cardiac cycle through the spontaneous generation of action potentials.
   (b) **Atrioventricular node**: Delays the impulse.
   (c) **Bundle of His**: Distributes the action potentials over the ventricles (resulting in ventricular contraction).
   (d) **Intercalated discs:**

2. (a) **Myogenic**: The heart muscle is capable of rhythmic contraction independently of any external nervous stimulation.
   (b) **Evidence**: When the heart is removed from its nervous supply (if provided with adequate oxygen, ions, and fluids) will continue to beat.

3. Delaying the impulse at the AVN allows time for atrial contraction to finish before the ventricles contract.

4. (a) Increases heart rate via release of noradrenaline.
   (b) Decreases heart rate via release of acetylcholine.

5. (a) Increased arterial flow (in aorta and carotid arteries).
   (b) Stretch receptors in carotid sinus and aorta detect an increase in arterial flow and send afferent impulses to the inhibitory centre in the medulla. The inhibitory centre mediates a decrease in heart rate via the vagus nerve (the carotid and aortic reflexes). There is a subsequent decrease in cardiac output and arterial blood pressure.

6. (a) and (b) any of the following in any order:
   - Aortic pressure receptors (baroreceptors) in the wall of the aortic arch respond to increased arterial blood flow in the aorta.
   - Carotid baroreceptors in the carotid sinus respond to increased arterial flow in the carotid artery.
   - Baroreceptors in the vena cava and the right atrium respond to increased venous return (mediates an increase in heart rate - the Bainbridge reflex).

7. (a) Individuals have different tolerances to the same level of a substance (also depends on their previous intakes of other substances such as coffee). Different body mass will affect the response also.
   (b) Controls should drink an energy drink that does not contain guarana.

Diagnosing Heart Problems (page 221)

1. (a) Symptoms vary but could include: chest pain or discomfort, shortness of breath, nausea, sweating, feeling faint, pain in the shoulders back or jaw.
   (b) The physical symptoms of some heart abnormalities can be very similar, making a conclusive diagnosis difficult. Each heart abnormality has a distinctive ECG pattern, so ECG is used to provide a definitive diagnosis.

2. A normal ECG includes three positive deflections (P, R, and T, with the second being larger) interspersed by two negative deflections (Q and S with the second being the larger). Myocardial infarction shows a larger Q wave, and an ill-defined QRS complex. Angina shows ill-defined P and T waves, little Q wave deflection and an ill-defined S wave.

Control of Breathing (page 222)

1. The basic rhythm of breathing is controlled by the respiratory centre in the medulla which sends rhythmic impulses to the intercostal muscles and diaphragm to bring about normal breathing.

2. (a) Phrenic nerve: Innervates the diaphragm (which contracts and moves down in inspiration).
   (b) Intercostal nerves: Innervate the intercostal muscles (internal and external intercostal nerves and muscles) to bring about ribcage movements.
   (c) Vagus nerve: Sensory portion carries impulses from stretch receptors in the bronchioles to the respiratory center to inhibit inspiration.
   (d) Inflation reflex (also known as the Hering-Breuer reflex): The inhibition of the inspiratory center to end the breath in. Note: Sensory impulses from the stretch receptors in the bronchioles travel (via the vagus) to inhibit the inspiratory center and expiration follows. When the lungs deflate, the stretch receptors are not stimulated and the inhibition of the inspiratory center stops.

3. (a) Low blood pH increases rate and depth of breathing.
   (b) Sensory information from aortic and carotid chemoreceptors is sent to the respiratory center, which mediates the increase in breathing rate. Note: Sensory impulses are sent from the carotid bodies (chemoreceptors) via the carotid sinus nerve and then the glossopharyngeal nerve. Sensory impulses from the aortic bodies (chemoreceptors) travel in the vagus nerve. Low blood pH also stimulates the chemosensitive area in the medulla directly.
   (c) Blood pH is a good indicator of high carbon dioxide levels (and therefore a need to increase respiratory rate to remove the CO2 and obtain more oxygen).

Exercise and Blood Flow (page 223)

1. Answers for missing values are listed from top to bottom under the appropriate heading:

	At rest (% of total)	Exercise (% of total)
Heart	4.0	4.2
Lung	2.0	1.1
Kidneys	22.0	3.4
Liver	27.0	3.4
Muscle	15.0	70.2
Bone	5.0	1.4
Skin	6.0	10.7
Thyroid	1.0	0.3
Adrenals	0.5	0.1
Other	3.5	1.0

2. The heart beats faster and harder to increase the volume of blood pumped per beat and the number of beats per minute (increased blood flow).

3. (a) Blood flow increases approximately 3.5 times.
   (b) Working tissues require more oxygen and nutrients than can be delivered by a resting rate of blood flow. Therefore the rate of blood flow (delivery to the tissues) must increase during exercise.

4. (a) Thyroid and adrenal glands, as well as the tissues other than those defined in the table, show no change in absolute rate of blood flow.
   (b) This is because they are not involved in exercise and do not require an increased blood flow. However, they do need to maintain their usual blood supply and cannot tolerate an absolute decline.

5. (a) Skeletal muscles (increases 16.7X), skin (increases 6.3X), and heart (increases 3.7X)
   (b) These tissues and organs are all directly involved in the exercise process and need a greater rate of supply of oxygen and nutrients. Skeletal muscles move the body, the heart must pump a greater volume of blood at a greater rate and the skin must help cool the body to maintain core temperature.

6. Heart size increases because (like any muscle) it gets bigger with work. The larger size also means it pumps a greater volume of blood more efficiently.

7. Endurance athletes have a smaller body weight.

8. With each stroke, the heart pumps a larger volume of blood. Less energy is expended in pumping the same volume of blood.

9. A lower resting heart rate means that for most of the time, the heart is not working as hard as in someone with a higher resting heart rate.

Measuring Lung Function (page 225)

1. (a) Taller people generally have larger lung volumes and capacities.
   (b) Males have larger lung volumes and capacities than females.
   (c) After adulthood, lung volume and capacity declines with age. Children have smaller lung volumes and lung capacities than adults.

2. (a) Forced volume is a more useful indicator of impairment of lung function than a tidal volume because people use only a small proportion of their lung volume in normal breathing.
   (b) Spirometry can be used to measure the extent of recovery of lung function after treatment.

3. (a) Tidal volume   vol: $0.5 \, dm^3$
   (b) Expiratory reserve volume   vol: $1.0 \, dm^3$
   (c) Residual volume   vol: $1.2 \, dm^3$
   (d) Inspiratory capacity   vol: $3.8 \, dm^3$
   (e) Vital capacity   vol: $4.8 \, dm^3$
   (f) Total lung capacity   vol: $6.0 \, dm^3$

4. **G**: Tidal volume is increasing as a result of exercise.

5. PV: $15 \times 0.4 = $ or $6 \, dm^3$

6. (a) During strenuous exercise, PV increases markedly.
   (b) Increased PV is achieved as a result of an increase in both breathing rate and tidal volume.

7. (a) There is 90X more CO_2 in exhaled air than in inhaled air ($3.6 \div 0.04$).
   (b) The CO_2 is the product of cellular respiration in the tissues. **Note**: Some texts give a value of 4.0% for exhaled air (100X the CO_2 content of inhaled air).
   (c) The dead space air is not involved in gas exchange therefore retains a higher oxygen content than the air that leaves the alveoli air. This raises the oxygen content of the expired air.

Exercise and Health (page 227)

1. (a) Blood flow rate is increased during exercise through increased rate and force of heart contraction and skeletal muscle contraction, and dilation of blood vessels.
   (b) Physiological effects of regular exercise include: An increase in muscular strength and flexibility, more efficient heart function, improvement in immune function, increased concentration, and higher energy levels.
   (c) These changes in physiology equip the body to handle the usual, everyday events of life with less effort and stress. Ongoing benefits include improved blood flow and more efficient organ function, faster tissue repair, better weight control, improved sleep and stress management, and better immune function. **Note**: These benefits do not extend to over-training, when stresses exceed the health benefits gained from exercise.

2. (a) **Increase in lean muscle and decreased body fat**:
   Health benefits: General increase in level of fitness and muscular performance with the various physiological and psychological benefits that stem from this.
   Physiological mechanism: An increase in the amount of muscle tissue that can be recruited into work and a decrease in the amount of non-active tissue (fat) that must be moved around.
   (b) **Increased muscular strength and endurance**:
   Health benefits: Improved stamina and less chance of injury when exercising.
   Physiological mechanism: Improvements in cardiovascular and ventilation performance and greater resilience during activity.

3. (a) The Harvard Alumni Health Study was restricted to male Harvard graduates who began their studies between 1916 and 1950 and were still living in 1966. Therefore, it had a very narrow catchment for participation, and a low diversity of participants.
   (b) The study carried a high level of bias because the subjects were all males, highly educated, middle aged, and predominantly white. Most would be of a high socio-economic group and most likely highly motivated. They would have access to good health care and quality of food, so many socioeconomic factors contributing to coronary heart disease were not contributing factors to this study. The results are only useful for the small demographic represented in the study, and could not be reasonably applied to the wider population.

Overtraining and Injury (page 229)

1. Training causes damage to the body's systems, especially muscles. During rest, these systems are repaired. The rest period must be sufficient for the body to repair the damage sustained during training.

2. During recovery, the repair carried out on the muscle goes beyond the original level of damage. This process (**supercompensation**) increases muscle mass and the body's capacity to do work.

3. (a) Overtraining occurs when the recovery time between training sessions is too short and the body does not have time to fully repair itself.
   (b) Overtraining causes a range of physical and psychological effects. For example the body becomes physically tired, lacks coordination under stress, and has reduced capacity for speed and endurance. People can become irritable and lose enthusiasm for their sport.
   (c) Overtraining fatigues the body. It does not fully recover or repair itself between sessions and is under general stress. Injuries are more likely to occur under these conditions.

Correcting Sports Injuries (page 230)

1. Applying the RICE principle helps to reduce swelling and inflammation at the injury site, both of which cause blood vessel constriction and reduce blood flow to the injury site. Maintaining a good blood flow to the site of injury reduces pain and stiffness, and aids the recovery process. Rest allows the injury time to recover and repair itself.

2. Keyhole surgery uses only small incisions to access the surgical site. This makes it less invasive and results in minimal damage to the body. The body heals far more quickly than from open surgery, which uses a larger incision. Keyhole surgery also produces less scar tissue.

3. Some prosthetics give athletes a competitive advantage over able bodied athletes (e.g. their light weight material and the superior forces produced propel them further and faster). They are regarded as a technological aid not available to all.

Performance Enhancing Drugs (page 231)

1. The pressure to win or improve a performance is often an important factor, especially in professional sports where an improved performance or a winning result may produce a renewed contract for the athlete, more money, or sponsorship.

2. Banned substances must meet 2 of 3 criteria: 1) is it harmful? 2) does it enhance performance? 3) is its use fair to other athletes who are not using it? However some substances (e.g. creatine and caffeine) are not banned even though they may be harmful and/or enhance performance. Caffeine is found in many products and its intake could be difficult to avoid, so sporting bodies permit its use because it would be too difficult to enforce its ban. Users of caffeine supplements get an advantage that non-users do not and they are not penalised.

Key Terms: Mix and Match (page 232)

Atrioventricular node (M), Cardiac output (Q), Causation (R), Correlation (L), ECG (J), Exercise (D), Injury (O), Heart (U), Heart rate (P), Homeostasis (A), Hormone (S), Hypothalamus (F), Keyhole surgery (L), Myogenic (H), Negative feedback (V), Overtraining (T), Performance enhancing substance (W), Prostheses (B), Sinoatrial node (N), Spirometer (G), Thermoregulation (C), Tidal volume (K), Ventilation (E).

Plant Responses (page 234)

1. Light (including the light/dark cycle), gravity, temperature, touch, chemicals.

2. Appropriate responses to environmental stimuli enable the plant to synchronise its daily cycles and seasonally important events, such as germination, with environmental cues. Appropriate responses enhance survival in different environments.

Investigating Phototropism (page 235)

1. (a) Auxin
   (b) Positive phototropism
   (c) Point A: Cells remain small
       Point B: Cells elongate (become longer)
   (d) Side B
   (e)

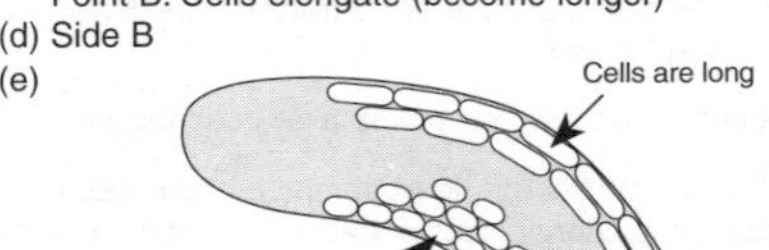

2. The action of light falling on the shoot tip is responsible for initiating the phototropic response. If the results of the experiments outlined in the lower panel (Q3) of this page were also known, one could surmise that the hormone bringing about the response was synthesised just below the region of elongation and transported there, where its uneven distribution either side of the shoot tip caused bending.

3. Plant A: Exhibits phototropism, bends towards sun.
   Plant B: No phototropism, does not bend.
   (Auxin synthesised in the region below the shoot tip and transported there).

Investigating Gravitropism (page 236)

1. (a) In shoots, more auxin accumulates on the lower side of the shoot. In response to higher auxin levels here the cells on the lower side of the stem elongate and the shoot tip turns up.
   (b) In roots, the accumulation of auxin on the lower side inhibits elongation (since this is the response of roots to high auxin). The cells on the upper side therefore elongate more than those on the lower side and the root tip turns down.

2. (a) Approx. 10^{-3} mg l^{-1} (b) Stem growth is promoted.

3. (a) Stems: A negative gravitropic response ensures shoots to turn upwards towards the light (important when light may be absent as it is deep in the soil).

(b) Roots: Positive gravitropism ensures roots turn down into the soil so that they can begin obtaining the water and minerals required for growth.

Photoperiodism in Plants (page 237)

1. Light induced responses involving phytochrome (a and b any two of): Seed germination, leaf growth, flowering (long day plants), chlorophyll synthesis.

2. **Phytochrome** is a photosensitive pigment which changes from the inactive form to the active form on absorption of light. The plant measures daylength by the amount of phytochrome of each form present.

3. (a) Day length (perhaps more importantly, night length).
   (b) Any one of: (1) Flowering at the same time ensures that other flowers will be available to provide/receive pollen. (2) Synchronisation with periods of high insect activity may assist pollination.

4. Environmental cues for triggering and breaking **dormancy** include: Decreasing day length (short days); long, cold nights; dry, nitrogen deficient soils. The necessity of exposure to cold to break dormancy ensures the plant does not "mistake' transient environmental cues.

5. **Dormancy** (arrested growth and metabolism) occurs when conditions for plant growth and survival are poor (such as occurs in the winters of temperate regions). Dormancy allows survival over the inclement period. **Vernalisation** (the low temperature stimulation of flowering) is adaptive in temperate regions because, after a relatively long period of intense cold, a period of warmth indicates that the winter has passed and the new growing period has begun.

6. (a) Short-day plants: Flower only when the day length is short (average: 10 h).
   (b) Long-day plants: Flower only when the day length exceeds a certain minimum value (average: 14 h).

7. Short-day plants are really long night plants, requiring a night length of more than a minimum value. In the experiment outlined, a short-day plant failed to flower when a long night was interrupted by a short period of light (the plant interpreted this as a short night irrespective of the short day prior to it).

Response to Temperature in Plants (page 239)

1. (a) Dormancy: A condition induced by cold and/or shorter days and typified by extremely low metabolic rate and a suspension of growth and development.
   (b) Flowering: Some plants will only flower after the plant has been exposed to a period of low temperatures.
   (c) Germination: The seeds of some plants will only germinate after exposure to the appropriate period of low temperature (stratification). In some Australian species, the seeds must be exposed to the intense heat of a bushfire before germinating.

2. Vernalisation ensures the plant will flower during the warm weather (springtime) when the chances of successful pollination and fruit set are higher. The necessity of a **prolonged** exposure to cold ensures the plant does not "mistake' transient environmental cues.

3. Evergreens already have leaves ready to photosynthesise and channel energy into seed production during a short growing season. Their lower productivity in these regions is not sufficient to compensate for leaf loss and they must retain their leaves. In contrast, deciduous trees take much longer to regain their leafed status before producing flowers.

4. (a) Over a twelve hour period the tulip flower will open in the morning and close in the afternoon (and remain closed until the following morning).
   (b) Closing the flower during the night reduces water loss as well as reducing the likelihood that the nectary will be plundered by non-pollinating insects.

Auxins, Gibberellins, and ABA (page 240)

1. **Auxins** in the growing leaves of the apical bud are synthesised in concentrations high enough to suppress the growth of the buds below. Consequently, the main shoot grows more vigorously than the lateral shoots.

2. **Gibberellins** are strong promoters of elongation in stems by stimulating both cell division and cell elongation. In seeds, gibberellins are responsible for breaking dormancy and stimulating the growth of the embryo and emergence of the seedling.

3. In response to water stress, ABA stimulates the closing of stomata. **Note**: It does this by stimulating the loss of K^+ from the guard cells.

Plant Hormones and Applications (page 241)

1. (a) **Gibberellins**: any one of the following:
   Flowering: Stimulates flowering in some plants.
   Stem: Cell division and elongation. Can cause spectacular stem elongation in some plants.
   Seeds: Breaks seed dormancy in many plants, and can break winter dormancy of plant.
   General: Delays onset of senescence.
   (b) **Cytokinins**: any one of the following:
   General: Promotes cell division and differentiation. Promotes lateral bud development. Keeps shoot and root growth in balance. Inhibits leaf abscission and senescence. Sprayed on cut flowers it stops ageing. Involved in seed germination in some plants.
   (c) **Ethene (ethylene)**: any one of the following:
   Abscission: Promotes abscission (dropping off) of leaves, flowers, fruits.
   Fruit: Promotes fruit ripening.
   (d) **Abscisic acid**: any one of the following:
   Leaf stomata: Causes stomatal closing in response to water stress, limiting water loss.
   Abscission: Promotes leaf ageing and abscission.
   Seeds: Induces dormancy in seeds.

2. (a) The growing leaves of the apical bud synthesise auxin in concentrations high enough to suppress the growth of the buds below. Consequently, the main shoot grows more vigorously than the lateral shoots.
   (b) Auxin is a strong promoter of stem elongation and promotes the differentiation of stem tissue.
   (c) Auxin is transported from the tips of the stems and roots to the vascular cambium at other regions of the plant where it promotes cell enlargement and differentiation.

3. Pruning removes the primary source of auxin synthesis and so also the suppressing effect of the apical shoot.

Consequently, the lateral shoot growth increases.

4. (a) Auxins (IAA), in the form of rooting powders, applied to cuttings will promote root growth (controls the differentiation of tissues).
   (b) Ethene levels in a storage facility can be raised to induce fruit ripening.
   (c) Gibberellins can be applied to break the dormancy of seeds and buds and promote germination.
   (d) Application of abscisic acid (ABA) to seeds will promote seed dormancy. Note that no single substance is involved in inducing dormancy: rather the dormancy is controlled by the interaction of growth inhibitors (ABA) and growth promoters (gibberellins, cytokinins, and ethene).

5. (a) Auxin (IAA), gibberellins
   (b) Ethene, abscisic acid (ABA)

Detecting Changing States (page 243)

1. Appropriate responses aid survival and enable effective interpretation about the changing state of the environment. For example, pain withdrawal response prevents a burned hand, running away from a snarling dog avoids attack etc.

2. (a) and (b) any from the examples provided (or others). Answers provided as stimulus (receptor):
   (a) <u>External stimuli and their receptors</u>: Light (photoreceptor cells); gravity (hair cells/vestibule of the inner ear); sound/vibration (hair cells in the cochlea of the inner ear); airborne chemicals (olfactory receptors in the nose); dissolved chemicals (taste buds); external pressure (dermal pressure receptors, e.g. Pacinian corpuscle, Meissner's corpuscles); pain (nerve endings in the dermis); temperature (nerve endings in the skin).
   (b) <u>Internal stimuli and their receptors</u>: Blood pH/ carbon dioxide level (chemoreceptors in blood vessels); blood pressure (baroreceptors); stretch (proprioreceptors, e.g. muscle spindle).

Taxes and Kineses (page 244)

1. A **kinesis** describes movement of a cell or organism in which the *rate* of movement depends upon the intensity (rather than direction) of the stimulus. An example is the increased activity of body lice when temperature increases over 30°C. In contrast, a **taxis** is a *directional* movement in response to an external stimulus. An example is the negative phototaxis of maggots.

2. Such simple behaviours allow an animal to continue making random orientation responses until it is in an environment that is favourable to its survival (e.g. body lice must reach an appropriate host, as indicated by the surface skin temperature). Once positioned, the behaviour will cease until environmental conditions change once again.

3. (a) Gravi –   Gravity
   (b) Hydro –   Water/moisture/humidity
   (c) Thigmo –   Touch
   (d) Photo –   Light
   (e) Chemo –   Chemical
   (f) Thermo –   Heat

4. (a) Snails: Negative gravitaxis
   (b) Moth: Positive chemotaxis

(c) Louse: Kinesis in response to temperature
(d) Lobster: Positive thigmotaxis
(e) Mosquitoes: Positive thermotaxis
(f) Maggots: Negative phototaxis

5. 30°C

Pheromones (page 245)

1. (a) **Hormones** are chemicals produced by an animal in endocrine tissue and released into the bloodstream where they influence the physiology of target tissues **within the body**. In contrast, **pheromones** are chemicals produced by an animal and released into **the environment** where they influence the behaviour or physiology of other individuals of the same species, e.g. as a sexual attractant.
   (b) Pheromones are species specific so that the message is received by 'the right organism'. Example included: To attract a mate and trigger specific courtship behaviours, or to guide other individuals to food.

2. A pheromone trap (baited with synthetic pheromone) would attract the male insect pests (females release the sex attractant) and contain them for later disposal. The fewer males remaining to breed would help decrease the overall size of the pest population.

3. Roles of pheromones:
   - As mate attractants, e.g. moths.
   - In chemical trails to locate food sources, e.g. ants.
   - As an alarm to warn of danger, e.g. ants.
   - To maintain social order, e.g. honeybee colonies.
   - To chemically sample the environment, e.g. mammals and reptiles.

Reflexes (page 246)

1. Higher reasoning is not a preferable feature of reflexes because it would slow down the response time. The adaptive value of reflexes is in allowing a very rapid response to a stimulus.

2. A spinal reflex involves integration within the spinal cord, e.g. knee jerk (monosynaptic) or pain withdrawal (polysynaptic). A cranial reflex involves integration within the brain stem (e.g. pupil reflex).

3. (a) A monosynaptic reflex arc involves just two neurones and one synapse (e.g. knee jerk reflex) and a polysynaptic reflex arc involves two synapses through a relay or interneurone, e.g. pain withdrawal.
   (b) A monosynaptic reflex arc, because there is one less synapse over which diffusion of neurotransmitter must occur.

4. (a) Newborn reflexes equip them with the appropriate survival behaviour in their otherwise helpless state: the rooting reflex helps them to locate a nipple, the suckling reflex insures feeding, the startle reflex induces crying, invoking a parental care response, the grasp reflex ensures they keep contact with (usually) the mother.
   (b) The presence of these reflexes indicates appropriate development. An absence of reflex behaviours in newborns may indicate neural damage or developmental impairment.

Nervous Regulatory Systems (page 247)

1. (a) The **sensory receptors** receive sensory information (information about the environment) and respond by generating an electrical response (message).
   (b) The **central nervous system** (CNS) processes the sensory input and coordinates an appropriate response (through motor output).
   (c) A system of **effectors** bring about an appropriate response.
   Note: Together these systems function to bring about appropriate (adaptive) responses to the environment so that homeostasis (steady state) is maintained.

2. (a) and (b) any two in any order:
   - Nervous control involves transmission across synapses, hormonal control involves transport of chemicals in the blood.
   - Nervous control is rapid, hormonal control is slower.
   - Nervous control acts in the short term and its effects are short lived, hormonal control is longer acting.
   - Nervous control is direct and through specific pathways, hormonal control is widespread, affecting target cells throughout the body (although these may be quite specific).
   - Nervous control causes muscular action directly, hormonal control generally acts by changing metabolic activity.

Neurone Structure and Function (page 248)

1. Missing panels as follows:
   <u>Length of fibres</u>:
   Sensory neuron: Long dendrites, short axons
   Motor neuron: Short dendrites, long axons
   <u>Function</u>:
   Sensory neuron: Conducts impulse to CNS
   Motor neuron: Conducts impulse to effector (muscle or gland)

2. (a) Myelination increases the speed of impulse conduction.
   (b) Oligodendrocytes
   (c) Schwann cells
   (d) Neurones in the PNS frequently have to transmit over long distances so speed of impulse conduction is critical to efficient function.

3. Myelin prevents ions from entering or leaving the axon along myelinated segments and so prevents leakage of charge across the neurone membrane. The current is carried in the cytoplasm so that the action potential at one node (gap in the sheath) is sufficient to trigger an action potential at the next. Myelin has another advantage too in reducing energy expenditure since fewer ions overall need to be pumped to restore resting potential after an action potential has passed.

4. (a) Faster conduction speeds enable more rapid responses to stimuli.
   (b) Increasing axon diameter increases conduction velocity because it reduces the longitudinal resistance, so local currents can spread further.

Chemical Synapses (page 250)

1. A **synapse** is a junction between the end of one axon and the dendrite or cell body of a receiving neurone. **Note**: A synapse can also occur between the end of one axon and a muscle cell (neuromuscular junction).

2. Arrival of a nerve impulse at the end of the axon causes an influx of calcium. This induces the vesicles to release their neurotransmitter into the cleft.

3. Delay at the synapse is caused by the time it takes for the neurotransmitter to diffuse across the gap (synaptic cleft) between neurones.

4. (a) Neurotransmitter (NT) is degraded into component molecules by enzyme activity on the membrane of the receiving neurone (in this case, acetylcholinesterase acts on Ach to produce acetyl and choline). This reference is to the cholinergic synapse pictured. Continued action of the NT at adrenergic synapses is prevented by reuptake of the NT (noradrenaline) by the presynaptic neurone.
   (b) The neurotransmitter must be deactivated so that it does not continue to stimulate the receiving neurone (continued stimulation would lead to depletion of neurotransmitter and fatigue of the nerve). Deactivation allows recovery of the neurone so that it can respond to further impulses.

5. Function of acetylcholine (either of):
   - It occurs at the synapse between motor neurones and muscle cells, where it causes depolarisation of the muscle cell (and leads to muscle contraction).
   - Ach is released by all parasympathetic fibres (autonomic nervous system) where it may cause an excitatory or inhibitory response (depending on the postsynaptic receptor with which it interacts). NOTE: Ach is also released by preganglionic fibres in the sympathetic division of the ANS.

6. The amount of neurotransmitter released influences the response of the receiving cell (response strength is proportional to amount of neurotransmitter released).

Action Potentials (page 251)

1. An action potential is a self-regenerating depolarisation (electrochemical signal) that allows excitable cells (such as muscle and nerve cells) to carry a signal over a (varying) distance.

2. (a) The ability to conduct electrical impulses.
   (b) They cannot conduct electrical impulses.

3. An **action potential** passes along a nerve because the depolarisation occurring in one region of the axon makes the next region of the axon more permeable to sodium ions (and more likely to conduct the impulse).

4. Because the refractory period makes the neurone unable to respond for a brief period after an action potential has passed, the impulse can pass in only one direction along the nerve (away from the cell body).

5. The nervous system interprets nerve impulses correctly because it records where they have come from and "knows" where they go to. Different regions of the brain are responsible for sorting out, interpreting, and integrating the nerve impulses from different sources.

6. NOTE: Attentive students will notice that this trace does not perfectly match the idealised schematic of an action potential (namely the hyperpolarisation does not fall below threshold and resting potential is more positive than the starting resting potential). Differences in electrophysiological techniques used in recording

lead to departures from the schematic. In this case, the real action potential is sitting atop a voltage offset caused by the current pulse used to stimulate the axon.

(a) B: Membrane depolarisation
D: Hyperpolarisation
E: Return to resting potential
C: Repolarisation
A: The membrane's resting potential
(b) Point 1: Stimulus causes opening of Na^+ channels and Na^+ begins to flow into the cell.
(c) Rapid Na^+ influx = rapid depolarisation.

The Structure of the Eye (page 253)

1. (a) **Cornea**: Responsible for most of the refraction (bending) of the incoming light.
(b) **Ciliary body**: Secretes the aqueous humour which helps to maintain the shape of the eye and assists in refraction.
(c) **Iris**: Regulates the amount of light entering the eye for vision in bright and dim light.

2. (a) The incoming light is refracted (primarily by the cornea) and the amount entering the eye is regulated by constriction of the pupil. The degree of refraction is adjusted through **accommodation** (changes to the shape of the lens) so that a sharp image is formed on the retina.
(b) Accommodation is achieved by the action of the ciliary muscles pulling on the elastic lens and changing its shape. **Note**: When the ciliary muscle contracts there is decreased pressure on the suspensory ligament and the lens becomes more convex (to focus on near objects). When the ciliary muscle relaxes there is increased tension on the suspensory ligament and the lens is pulled into a thinner shape (to focus on distant objects).

3. (a) The pupil is the hole through which light enters the eye. By constricting (in bright light), the pupil can narrow the diameter of this entry point and prevent light rays entering from the periphery. In dim light, the opposite happens; the pupil expands to allow more light into the eye.
(b) Control of over the entry of light is appropriate as a reflex activity because the response is immediate and unconscious. In this way, the eye can be protected against damage and vision optimised without the need to consider the action necessary (which would be slow and inefficient).

4. (a) Image (point of focus) is in front of the retina.
(b) Image (point of focus) is behind the retina.

5. (a) Myopia: Concave lens diverges the incoming light rays so that they have to travel further through the eyeball and are focused directly on the retina.
(b) Hypermetropia: Convex lens converges incoming light so that the image falls directly on the retina.

The Physiology of Vision (page 255)

1. (a) The retina is the layer of light sensitive tissue lining the inner surface of the eye. It comprises several layers of neurones, including the photoreceptor rods and cones, interconnected by synapses. Light induces the production of neural signals in the rods and cones, which are processed by other neurones of the retina.
(b) The optic nerve is formed from the axons of the retinal ganglion cells, which carry the processed neural signals, the form of action potentials, from the retina through the optic chiasma to the visual cortex in the cerebrum. It also contains the blood vessels supplying the retina.

2. Visual acuity is greatest in the central fovea as this is the area of greatest density of cones; the photoreceptor cells specialised for high visual acuity, as well as colour vision. Rods are absent in the central fovea.

3.

Feature	Rod cells	Cone cells
Visual pigment(s):	Rhodopsin (no colour vision)	Iodopsin (three types)
Visual acuity:	Low	High
Overall function:	Vision in dim light, high sensitivity	Colour vision, vision in bright light

4. (a)-(c) in any order
(a) **Photoreceptor cells (rods and cones)** respond to light by producing generator potentials.
(b) **Bipolar neurones** form synapses with the rods and cones and transmit the changes in membrane potential to the ganglion cells. Each cone synapses with one bipolar (=high acuity) whereas each rod synapses with many bipolar cells (=high sensitivity).
(c) **Ganglion cells** synapse with the bipolar cells and respond with depolarisation and generation of action potentials. Their axons form the optic nerve.

5. Each **rod** cell synapses with a number of bipolar cells and this gives poor acuity but high sensitivity. Each **cone** cell synapses with only one bipolar cell and this gives high acuity but poor sensitivity.

6. (a) A photochemical pigment molecule contained in the membranes of the photoreceptor cells that undergoes a structural change when exposed to light (and is therefore light-sensitive).
(b) Rhodopsin.
(c) Opsin (protein) + retinal (Cofactor).
(d) Light alters the shape of the rhodopsin and it breaks into its components (opsin + retinal). Light activation alters the rhodopsin from the *cis* form into the *trans* form.

7. Light falling on the retina causes structural changes in the photopigments in the rods and cones. These changes lead to the development of graded electrical signals (hyperpolarisations) which spread from the rods and cones, via the bipolar neurones, to the ganglion cells. The ganglion cells respond by depolarisation and transmit action potentials to the brain.

The Human Brain (page 257)

1. (a) Breathing/heartbeat: brainstem (medulla)
(b) Memory/emotion: cerebrum
(c) Posture/balance: cerebellum
(d) Autonomic functions: hypothalamus
(e) Visual processing: occipital lobe
(f) Body temperature: hypothalamus
(g) Language: motor and sensory speech areas

Imaging the Brain (page 258)

1. (a) X-ray images of the body are taken from multiple angles and in multiple slices. Computer software generates a 3-D image of the body. CT scans work best on dense material such as bone, but iodine-based contrast agents can be used to enhance some tissue (e.g. blood vessels).
   (b) The patient is exposed to high energy x-rays (radiation) which may increase the risk of the patient developing cancer. The scan time is limited to about one second to minimise risk. **Note:** Some patients also have allergic reactions to the iodine substances used to improve the resolution of the image.

2. (a) MRI scans give better images of soft tissue than CT scans can. The patient is not exposed to any radiation during MRI scans.
   (b) fMRI maps the change in blood flow related to neural activity in the brain. fMRI can be used to identify which regions of the brain are affected in Alzheimer's patients. fMRI could be used to examine the effectiveness of future Alzheimer's treatments on brain activity.
   (c) Patients with metal body parts can not be scanned because of the magnetic field.

Visual Development (page 259)

1. Hubel and Wiesel's experiments on the development of vision showed that if visual stimuli were impaired during certain critical developmental periods in young kittens, their vision did not develop normally. This is because specific neural connections were not made, and the development of the neural pathways did not proceed normally. In kittens, the critical window for visual development is between 4-8 weeks of age. If kittens were exposed to visual stimuli either side of this critical period, the effects of visual deprivation were less obvious.

2. From the graphs, impaired visual development started at approximately 23-26 days, but Hubel and Wiesel's wider experiments showed the critical window is between 4-8 weeks of age.

3. Kittens and monkeys share similarities in their visual systems with humans. Using animals with similar body systems to humans allows the results gathered from the studies to be applied to humans more readily than results from organisms with physiologically quite different systems.

4. Brain development models help researchers understand how the brain develops and functions, and to predict what developmental changes may occur under certain conditions. Abnormal development can be studied, and researchers can determine the effects of the abnormalities, and if these abnormalities can be corrected at a later stage. These types of studies can be used to help understand and treat developmental disorders, to develop and understand brain surgery techniques, to develop education systems (by understanding how and when learning occurs), or to improve therapies and treatments for specific conditions (e.g. autism, Alzheimer's).

5. There are both pros and cons of using animals for research. The benefits are:
   - Minimal harm to humans.

- The types of animals used are often small and can be kept in confined spaces so many experiments can be carried out.
- Some have short generation times, so results are obtained quickly.
- Results from animals with body and brain systems similar to humans can be extrapolated (with caution) to humans.
- Testing potentially dangerous medicines or procedures can be carried out on animals and once safe can be used on humans.

Arguments against animal testing include:
- The animals may suffer mental trauma (especially intelligent animals such as rats, cats, dogs, monkeys and apes).
- The animals may suffer physical trauma.
- The models/ drugs developed may not be usefully applied to humans due to different physiology.
- It is not morally right to subject animals to tests we ourselves would not undertake.

Nature versus Nurture (page 261)

1. Some examples showing that genes influence human development includes:
   - Studies from IQ tests show twins and biological siblings have a close correlation in the IQ scores while foster siblings do not.
   - Visual development follows a similar pattern in all infants (development of distance and spatial interpretation).
   - Alcoholism has a heritable component.

   Evidence that the environment influences our development includes:
   - fMRI scans showing adults from different cultures view scenes in different ways.
   - Infants can learn any language if exposed to it at a critical time period.
   - The brain can be fooled about size perception when viewing objects at different distances.

Habituation to Stimuli (page 262)

1. Habituation occurs when the body learns that a specific stimulus has no consequence. Changes in the brain and sensory systems stop the body from continually reacting to the stimulus.

2. Habituation is adaptive because it prevents continued responses to stimuli that provide no reinforcement (reward or punishment). This (1) prevents sensory overload and (2) enables animals to respond to the stimuli that are associated with a consequence.

Dopamine and Behaviour (page 264)

1. Behaviour is based on both genetics and the environment. Although genes may predispose a person to behaving in a certain way, the environment plays a major part in whether or not that behaviour is carried out. In addition, interactions between genes influence their final expression in the phenotype, so it becomes difficult to identify which gene is the 'cause' of the behaviour. Diet also has a major effect on behaviour (through physiological mechanisms) with many studies showing that artificial colourings and flavourings can affect behaviour in young people. Finally, human thought must be included. Humans can consciously

alter their behaviour in response to peer pressures and societal expectations (although again, both the environment and genetics play a part in this process).

2. Different dopamine levels are directly associated with certain behaviours. Low levels may cause people to become highly active. Dopamine levels increase when active and give the body a sense of wellbeing and satisfaction. In order to achieve this feeling, people with naturally low levels are often more active than normal as the body tries to produce more dopamine. Low levels are also associated with tendencies towards addiction, as the feelings gained from certain drugs are interpreted by the body as similar to dopamine and there is a physiological craving to maintain the feeling. High levels of dopamine are associated with fixation on certain objects and increased perception of the senses in a way that may cause people to hear and see things in an exaggerated way. Extremely high levels are associated with schizophrenia (a split between what is real and what is not) where the body begins to hear and see things that are not there.

3. Natural variations in the DRD4 gene affect its ability to produce receptors that bind with dopamine. Protein chains that are too long do not fold correctly and produce receptors that do not efficiently bind to dopamine. In this case the body will perceive the levels of dopamine to be low and act in a way to increase them (frequently by increasing activity).

4. Dopamine is present in a number of neural pathways. The link between the DRD4 gene and dopamine levels allows us to theorise that other neural pathways are affected by similar receptor genes. The link between a behaviour and a gene sets a precedence for finding other genes that affect behaviour. It lends weight to the theory that human behaviour is more influenced by genetics than by the environment.

Chemical Imbalances in the Brain (page 265)

1. Neurotransmitters are chemicals that transfer signals between neurones across a synapse (synaptic cleft). A neurotransmitter is released from the presynaptic neurone into the synaptic cleft where it interacts with a postsynaptic neurone to cause a response. The amount of neurotransmitter released influences the response of the receiving cell.

2. (a) Parkinson's disease is caused by a reduction in dopamine production due to the loss of nerve cells in the substantia nigra region of the brain. This reduces the stimulation in the motor cortex and results in slow physical movement and uncontrollable tremors.

   (b) Depression is caused by reduced serotonin released from the raphe nuclei resulting in reduced stimulation of neurones in the brain, but especially those related to emotion. People with depression often have feelings of low self esteem, guilt, regret, and suffer physical tiredness.

3. L-dopa is a precursor to dopamine, it can cross the blood-brain barrier where it is converted to dopamine. Dopamine stimulates the motor cortex, reducing the physical effects of Parkinson's. Dopamine can not be used as a treatment because it can not cross the blood-brain barrier, it therefore has no direct effect.

4. Antidepressants work by increasing serotonin levels in the brain. They have two main modes of action. SSRIs stop serotonin reabsorption by the presynaptic neurone (this causes increased serotonin levels and greater stimulation of the postsynaptic neurone). MAOIs prevent the breakdown of serotonin in the synaptic cleft once it has be released from the presynaptic neurone.

5. Ecstasy and SSRIs both increase relative serotonin levels. SSRIs prevent the presynaptic neurones reabsorbing serotonin from the synaptic cleft. Ecstasy works in two ways. It prevents serotonin re-uptake from the synaptic cleft, and also causes the serotonin transporters to work in reverse, flooding serotonin into the synaptic cleft and increasing its levels.

6. The huge spikes in serotonin secretion eventually lead to depletion of levels in the brain because the serotonin production can not keep pace with the increased secretion. This results in depression, psychiatric disorders, and memory problems.

Medical Applications of GE (page 267)

1. The details and example used in the of the student's discussion will vary but should include reference to the following:
   - Development of gene therapy techniques, including vectors and gene delivery techniques.
   - The use of microorganisms as biofactories to produce genetically engineered human proteins to treat specific human disorders (e.g. insulin for diabetes).
   - An understanding of how decoding the human genome has lead to the development of pharmacogenomics, and how this can be used to deliver specific tailor-made treatments to a patient. The advantages of such specific treatments options should be discussed.
   - How genetic modification can be used to develop new vaccines (e.g. development of the hepatitis B vaccine).

The Ethics of Genetic Modification (page 268)

1. This is the student's own discussion but some points for discussion are:
   – That the GM product and/or the GMO could have some unwanted harmful effect on humans or other organisms.
   – That the genetic modification would spread uncontrollably into other organisms (breeding populations of the same or different species).
   – Consumer choice is denied unless adequate labelling protocols are in place. If everything contains GM products, there is no consumer choice.
   – General fear of what is not understood (fear of real or imagined consequences).
   – Objections on the grounds that it is ethically and morally wrong to tamper with the genetic make-up of an organism.
   – Generation of monopolies where large companies control the rights to seed supplies & breeding stock.
   – Amongst plant GMOs, the indiscriminate spread of foreign genes.
   – Unusual physiological reactions e.g. allergies, to novel proteins.
   – Some animal rights issues may be justified if genetic modification causes impaired health.

The Human Genome Project (page 269)

1. **HGP**: Aims to map the entire base sequence of every chromosome in the human cell (our genome), to identify all genes in the sequence, determine what they express (protein produced), and determine the precise role of every gene on the chromosomes.
HGDP: The HGDP aims to map the differences in genomes between different racial and ethnic groups.

2. **Proteomics** is the study (including identification) of the protein products of identified genes. It relies on the knowledge gained by the HGP, but will ultimately provide the most useful information.

3. **Reactionary medicine** is most commonly used today, it treats a disease once symptoms are displayed It can be costly and ineffective. **Predictive medicine** studies an individuals genome and determines the likelihood of a specific disease developing. The disease can be treated, or in some cases, cured before it shows symptoms.

4. Student's own discussion. Suggestions for each issue listed in the table (pros and cons) are as follows:
   – Rights of third parties:
   (a) They should have the genetic information in order to make an informed decisions (about insurance premiums etc.) to the benefit of those with "favourable" genetic test results.
   (b) They should be denied the information because they could use it to unfairly discriminate against people with "unfavourable" genetic test results.

   – No treatment, therefore the knowledge is pointless:
   (a) Although there may be no treatment initially, treatment may become available and knowledge of genetic predisposition will allow informed decisions to be made at short notice if necessary.
   (b) Knowing that one has a disease and cannot do anything about it could create emotional problems for many people.

   – High costs of tests:
   (a) Although the costs are high, the knowledge is important to a person's health and to medical research generally and is justifiable.
   (b) If costs are not met by public funds, the high costs will preclude those individuals who cannot personally afford them.

   – Genetic information is hereditary:
   (a) Knowledge of an inherited disease or disorder lets family members assess their risks when planning their own lives.
   (b) Family members may feel forced to not have children if their risk of an inherited disorder is high.

Producing GMOs (page 271)

1. Organisms may be genetically modified by (1) the addition of a foreign gene, e.g. human insulin gene inserted into bacteria or yeast for the commercial production of human insulin. (2) Alteration of an existing gene so that a protein is expressed at a higher rate or in a different way. This GM technique is used in gene therapy. (3) Deletion or inactivation of an existing gene, e.g. the Flavr-Savr tomato which has had its ripening gene switched off. Gene inactivation also produces 'knock-out mice' which are used to study the physiological effects of particular genes.

2. The need to accelerate traditional breeding programmes and improve livestock and crops by increasing production and reducing susceptibility to diseases and pests. Also a desire to find new ways to produce proteins using plants and animals as biofactories.

Edible Vaccines (page 272)

1. (a) and (b) any of:
   – Plants producing the antigen can be grown locally (reducing costs of transport and distribution).
   – Syringes (and sterile conditions) are not required, saving costs and reducing the risk of infection.
   – Once established, edible vaccines can probably be produced at lower costs than conventional vaccines, and with lower consumer waste.

2. Any of: Immune response to antigens may be poor, foods suitable for producing vaccines may not be palatable or tolerated by the recipients.

3. Any fruit or vegetable that can be grown locally and eaten without preparation. Bananas are favoured being a tropical fruit and needing no washing or preparation.

4. A gene for antibiotic resistance serves as a marker. Cells lacking the marker will also lack the antigenic gene and can be killed off with antibiotic, thereby isolating the transgenic cells.

Human Proteins from GE Bacteria (page 273)

1. (a) High cost (extraction from tissue is expensive).
   (b) Non-human insulin (from pigs or cattle) is different enough from human insulin to cause side effects.
   (c) The extraction methods did not produce pure insulin so the insulin was often contaminated.

2. The insulin is synthesised as two (A and B) nucleotide sequences (corresponding to the two polypeptide chains) because a single sequence is too large to be inserted into the bacterial plasmid. Two shorter sequences are small enough to be inserted (separately) into bacterial plasmids.

3. The β-galactosidase gene in *E.coli* controls the transcription of genes, so the synthetic genes must be tied to that gene in order to be transcribed.

4. (a) Insertion of the gene: The yeast plasmid is larger and could accommodate the entire synthetic nucleotide sequence for the A and B chains as one uninterrupted sequence.
   (b) Secretion and purification: Yeast, being eukaryotic, has secretory pathways that are more similar to humans than those of a prokaryote. Secretion from the cell of the precursor insulin molecules is thus less problematic. Purification would be simplified because removal of β-galactosidase is not required.

5. Mass production of human proteins using GMOs facilitates a low cost, reliable supply for consumer use. The protein (e.g. insulin) is free of contaminants and, because it is a human protein, the side effects of its use are minimised.

6. In the future, gene therapy, where a faulty gene is corrected in the patient, could treat many inherited disorders of metabolism. The use of stem cells, which can differentiate and proliferate in the patient's tissue, may prove the best way to correct genetic disorders.

Key Terms: Crossword (page 275)

Answers Across	Answers Down
3. Postsynaptic	1. Parkinson's
5. Recreational	2. Neurotransmitter
6. Ldopa	4. MDMA
10. Human Genome Project	7. Drug
12. Ecstasy	8. Dopamine
13. Serotonin	9. Depression
14. Insulin	11. Synapse
15. Presynaptic	16. GMO
17. Genetic modification	